病翼驚秋枯形閱世消得

斜陽幾度餘音更苦甚

猶抱清商頓成悽楚漫想

薰風柳絲千萬縷

一九七六夏借王船山句書懷

徐逯光

T. C. Hsu

Springer-Verlag
New York
Heidelberg
Berlin

Human and Mammalian Cytogenetics

An Historical Perspective

With 27 illustrations

T. C. Hsu, Ph.D.
Section of Cell Biology
The University of Texas System Cancer Center
M.D. Anderson Hospital and Tumor Institute
Texas Medical Center
Houston, Texas 77030

QH
431
H755

Library of Congress Cataloging in Publication Data

Hsu, T C
  Human and mammalian cytogenetics.

  Bibliography: p.
  Includes index.
  1. Human genetics—History. 2. Mammals—Genetics—
History. 3. Cytogenetics—History. I. Title.
[DNLM: 1. Cytogenetics—History. QH605 H873h]
QH431.H755   575.2'1   79-14681

The use of general descriptive names, trade names, trademarks,
etc. in this publication, even if the former are not especially
identified, is not to be taken as a sign that such names, as
understood by the Trade Marks and Merchandise Marks Act,
may accordingly be used freely by anyone.

9 8 7 6 5 4 3 2 1

ISBN   0-387-90364-X   Springer-Verlag   New York

ISBN   3-540-90364-X   Springer-Verlag   Berlin   Heidelberg

Foreword

The history of science is mostly written retrospectively, a generation or two after the actual events being discussed. Science historians are now analyzing and evaluating the origins of evolutionary and genetical theory in the nineteenth century and a sort of "Darwin industry" seems to have grown up.

A history of mammalian cytogenetics by one of the main participants is, hence, a very welcome change, since it has a vividness, an immediacy and a personal flavor which these scholarly tomes and the official biographies of scientists mostly lack. The life of the author, Chinese-born, T. C. Hsu, has been a romantic and colorful one, and he is himself a unique personality, so that his book is a very unusual blend of reminiscences, history of his special field (which has transformed human genetics) and wise comments on the mistakes made along the way.

The best qualities of a very fine Chinese mind have contributed to Dr. Hsu's career, including this book. Those qualities (which seem to me especially Chinese) include a kind of transparent honesty, a very direct empirical approach to problems and superb technical ability. And precisely these qualities seem to have been needed for success in what is likely to be seen in the future as the golden age of human and mammalian chromosome studies. The story begins with the "Dark Age," before 1956, when cytological studies of sectioned human tissues were believed to support T. S. Painter's count of 48 chromosomes. The revolution which transformed cytogenetics was the work of many cytologists in Ameri-

can and European laboratories. An absolutely key discovery was that treatment of living cells with hypotonic
solutions would spread the chromosomes and greatly improve the quality of the preparations. A lucky accident in
Dr. Hsu's laboratory (a technician misread the scale and
prepared a hypotonic solution inadvertently) can be
compared with the *Penicillium*-contaminated petri plate
in Alexander Flemming's laboratory. In both cases a
prepared and alert mind seized on an anomalous result
that should not have occurred and converted it into a
major discovery.

The "hypotonic miracle," combined with other technical innovations, paved the way for the discovery by Jo
Hin Tjio and Albert Levan, in 1955, that the human
chromosome number is, in fact, 46 and not 48. The fascinating story of the later developments, with the discovery that mongolism is due to a specific anomaly of chromosome number 21 and that numerous other human
abnormalities are produced by visible alterations of the
chromosomes, should be read by every biologist. This is
emphatically a book which will inspire young readers: in
fact I predict that quite a number of the biologists of the
future will have been deflected from careers in other
fields by it.

Before becoming a human (and mammalian) chromosome cytologist, Dr. Hsu was an insect geneticist. I think
that his first scientific paper, published when he was still
in China, was on the chromosomes of a small Chinese
midge. When he came to the Texas Drosophila Laboratory in Austin early in 1948, he quickly made friends. But
it was apparent to all of us that he did indeed come from
some kind of bullock-wagon civilization very different
from ours. The first time I took him for a trip near Austin,
he saw the highway signs "U.S. 81" and concluded that
the speed limit was 81 miles an hour. That was barely
believable but when we turned onto U.S. 290 our new
graduate student looked distinctly alarmed! However he
soon overcame these initial misunderstandings and he
and I made a memorable trip to collect *Drosophila* and
grasshoppers all the way from Austin to Oregon. The flies
collected on that trip (in spite of flash floods, predatory
bears and other adventures) formed the basis for Hsu's
Ph.D. dissertation. But before that achievement there
were obstacles to be overcome—in particular the somewhat rigid ideas of Hsu's supervisor, the elderly Professor J. T. Patterson, who really didn't like Hsu's making any

discoveries which he had not himself predicted. On the trip Hsu had discovered a new species of *Drosophila* along the Yampa River in Colorado. He asked me to suggest a name for it, so on the spur of the moment I replied that it might perhaps be called *Drosophila yampa*. Patterson bristled with indignation. Stomping down the hall he muttered to me: "That Chinese boy's full of prunes! Thinks he's got a new species—well it isn't. And he wants to call it by an *unpronounceable Chinese name!*" (Patterson later named it *Drosophila flavomontana*).

After receiving his doctorate, Hsu went to work on cultured human and mammalian cells in the laboratory of Charles Pomerat in Galveston. By this time the communist regime had taken over in China and Lysenkoism seemed to be dominant there. So Hsu remained in the United States and was eventually joined by his wife and the daughter he had never seen.

There are some biologists who work with time-honored techniques but continue to turn out new discoveries. Others are technique innovators but never discover anything but techniques. Hsu is a technique innovator who has made fundamental discoveries in cytogenetics with the aid of those techniques. It is part of his charm that his work and his book are illuminated throughout with a very special kind of warm humanity. More than any other biologist I know, T. C. Hsu seems to *care*—for his experimental animals, for his colleagues and for humanity as a whole. I think that is why this history of modern mammalian cytogenetics is so different from any other I know in the field of cytogenetics. It will surely become a classic.

M. J. D. White
Canberra, Australia

Preface

In the early 1950s, soon after I accidentally discovered the hypotonic solution pretreatment method for studying human and mammalian chromosomes, I met Professor Franz Schrader, then the coeditor of *Chromosoma*. I asked him whether *Chromosoma* would be interested in publishing a paper on the technique, which I thought had great potential. "We'll take a paper presenting good data," said he, "and you may describe your technique in the Material and Methods section."

This terse conversation exemplified the prevailing attitude at that time regarding biological research and scientific contributions: Methodological and technological achievements were subordinate to fact-finding missions and conceptual advancements. After all, a Ph.D. meant a doctor of philosophy. How could a technical device be philosophical?

Things have changed, of course. Men have gone to the moon. Biologists (imagine, biologist!) talk about cyclic AMP, nucleotide reassociation kinetics, restriction enzymes, etc., as if these subjects were born with them. People recognize that a new technique may beget new sets of pertinent facts, which in turn may beget new concepts and new theories. In this little book, I hope to portray how human and mammalian cytology, a seemingly insignificant and innocuous plaything of some microscopists who had nothing better to do, blossomed from oblivion to prominence and entered many branches of contemporary biology and medicine. And all this depended on a few technical improvements!

A number of friends urged me from time to time to write a resumé of the development of the field of human and mammalian cytogenetics. At first I hesitated because I was (and still am) not familiar with all the facets of this area of research. But then I agreed to give it a try. Since I have actually witnessed the growth of this field and have shared the excitement and frustration with fellow workers, I suppose I am one of the persons qualified to summarize the history of this branch of biology. I thought I would relate some background information on the events that led to some of the discoveries and contributions (usually only told at cocktail hours), together with brief accounts of the achievements. I thought I would also express my own opinions occasionally, right or wrong. Therefore, this book is not really very scholastic. Nevertheless, I hope it may be interesting to read. However, because of my personal involvement, it is inevitable that I shall describe more of my own experiences than those of others. To them I offer my apology. I further apologize to friends who have made contributions to this field but whose names I fail to mention.

Customarily, authors like to quote a few lines from the Bible, verses from the literature, or utterances of famous persons in the past, to decorate their books or chapters. I choose to quote a few verses from an old Chinese poem as my frontispiece. The poem was composed in the late Sung Dynasty by Wang Yi-sun. Ostensibly, it describes a cicada who, singing in sobbing notes against the autumn winds, yearns for the sweet summer breeze sweeping through willow branches and bemoans the anticipation that he has not many sundowns to go. Of course the poet metaphorically lamented over his own impending age and his lack of fulfillment. It is not exactly a delight for a nonhistorian to write an historic account of a subject in which he played a role, for this only suggests that he, too, yearns for the sweet summer breeze!

Acknowledgments

In the course of preparing this book, many friends lent assistance in a variety of ways. Some provided me with information, others gave illustrative materials, and still others offered encouragement, criticisms, and comments on the draft. I would like to give my sincerest thanks to these: Drs. Ulfur Arnason, Frances E. Arrighi, Murray L. Barr, Hans Bauer, Kurt Benirschke, Larry L. Deaven, A. C. Fabergé, Patricia Farnes, Joseph G. Gall, David A. Hungerford, Harold P. Klinger, Jérôme

Lejeune, Albert Levan, C. C. Lin, Dan Lindsley, Sajiro Makino, Robert Matthey, Orlando J. Miller, Peter C. Nowell, Joseph T. Painter, Mary Lou Pardue, Sen Pathak, Rudolph Pfeiffer, T. T. Puck, Potu N. Rao, Margery W. Shaw, Elton Stubblefield, Jo Hin Tjio, and Jacqueline Wang-Peng. I also would like to thank my secretaries who typed and retyped the manuscript as I changed my mind: Mlles. Mary Bea Cline, Dorothy Holitzke, and Patricia Hall. A special gratitude is extended to Miss Joan McCay, who practically twisted my arms in urging me to write this book and who got me over my inertia.

T. C. Hsu

Contents

1 Introduction

The science of human and mammalian cytogenetics began very slowly. During the last two decades, however, it has made some remarkable progress, and increased activity shows no sign of abatement. For example, the arena for investigations on chromosomes has traditionally been the research laboratories, which tackled a variety of problems of fundamental importance. But today, investigations are also carried out in medical institutions as routine tools for diagnosis, prognosis, and counseling.

Probably few cytogeneticists would disagree that the history of human and mammalian cytogenetics can be divided roughly into four periods, separated by some significant events. The first period can be called the pre-hypotonic era, in which the use of classic sectioned or squash preparations of tissues *in situ* was the prevailing practice. I have written only a single chapter (Chapter 2) to review the activities of that era, since most of the data and conclusions of the period were later found to be erroneous or in need of revision.

The second period (Chapters 3–5) lasted seven years (1952–1959), including the rediscovery of the hypotonic solution pretreatment for cytological preparations, which turned out to be a key to the development of the field of human and mammalian cytogenetics. During this period, the diploid number of man, long believed to be 48, was corrected to 46. Characterization of the human chromosomes also began. In addition, some approaches developed by cell biologists, such as autoradiography and cell cloning, began to be utilized to study chromosomes.

Still, the activities during this period were minimal compared to those of the next two periods, which were stimulated by a number of factors—especially the discovery of trisomy (one or more triploid chromosomes) in man.

The third period (1959–1969) saw explosive activities in human cytogenetics and the beginning of research into mammalian chromosomes (Chapters 6–15). Needless to say, the finding that a clinical syndrome is intimately associated with an abnormal chromosomal constitution stimulated the progress of medical cytogenetics. In the meantime, the somatic cell hybridization system was established as a new tool to correlate genetics and cytology in higher animals.

The modern period (1969–onward) started with the invention of techniques to characterize individual metaphase chromosomes and to subdivide each chromosome into visually recognizable zones or bands. Without such techniques, much of the earlier work would have remained ambiguous or even erroneous. Biologists also began to combine molecular biology and cytology, and a new field, molecular cytogenetics, emerged. A cursory inspection of cytogenetic journals such as *Chromosoma* shows a heavy influx of articles in recent years dealing with molecular biology or molecular cytogenetics, whereas 10 or 15 years ago few could be found. Chapters 16–23 describe some of the events of this period.

The divisions of the history of human and mammalian cytogenetics into periods are marked by some landmark events and the dominant research activities that followed. This does not mean that research of the previous period ceased when a new period arrived. Nor does it mean that one period of research did not benefit from the research of the past. However, for convenience in describing a subject, I have chosen to present each subject in one chapter to avoid duplication. For example, somatic cell hybridization started early in the 1960s, and many contributions were made in that decade. But I present the subject in the modern period to avoid presenting it in two chapters.

The last chapter represents my personal views on what might come in the future. Naturally, these are all speculation, and the actual cause of development may be completely different.

It would be impractical to cite a large number of references because the bibliography would be enormous. However, I have cited some pertinent references mentioned in the text and have added a few review papers and books that are not referred to in the text but are excellent reading materials. References in the latter category are gathered as a list of suggested readings that precedes the cited references.

2 The Knights of the Dark Age

A few years ago, I was invited to participate in the Symposium on Biological Clocks, organized by the Argonne National Laboratory. Often before, my presentations had been arranged (logically) to follow papers on biochemistry and molecular biology, and that symposium was no exception. By way of an introduction, I commented that, with the presentations on biochemistry and molecular biology behind us, we were leaving the precise studies and entering the realm of ambiguity. I also commented that whereas molecular biology is beautiful—because of the imaginative ideas behind the research, the neat experimental protocol, the precise measurements of the results, and the meaningful interpretations—the beauty is primarily abstract, similar to that of many abstract paintings. No one gets much visual pleasure from a scintillation plot or a density gradient tracing. On the other hand, microscopic objects, particularly chromosomes, are themselves objects of beauty, analogous to the beauty of a Rembrandt.

My topic of that meeting was the kinetochore structure and the relationships between the kinetochore and the spindle microtubules, a collaborative work with my former colleague Bill Brinkley. I supplemented my presentation with many electron micrographs. Herbert Stern, the speaker after me, said, "Dr. Hsu compared his electron micrographs with Rembrandt paintings for visual enjoyment. I guess his electron micrographs and Rembrandt paintings have something in common: Lots of shade and not much light." He then proceeded to tell the audience that his Feulgen-fast green photos of plant chromosomes were more like Renoirs.

I tell this story of friendly jest to illustrate that chromosomes have attracted many microscopists not only because these sausage-like bodies represent vehicles of genetic material (and, hence, are biologically important) but also because they are hypnotically beautiful objects. Sometimes, after having obtained the information they sought, microscopists would continue to examine cytological preparations just for the enjoyment of looking at more mitotic or meiotic figures.

In the first half of this century, the development of cytogenetics relied heavily on studies of plant and insect chromosomes. Many important discoveries were made during examinations of these materials, and techniques were standardized. Yet the chromosome cytology of one of the most important groups of life forms, the vertebrates, was disproportionately underdeveloped. It is clear now that the great majority of data on vertebrate cytogenetics derived during this period were unreliable, and it is no exaggeration to call the first half of this century the Dark Age of mammalian and human cytogenetics.

During this period, there were two important technical improvements in chromosome cytology—the squash preparation methods and the colchicine pretreatment method—both invented by plant cytologists. Before Belling (1921) introduced the squash technique, cytologists used the paraffin section method to obtain their preparations. For studies of tissue organization (histology, embryology, and pathology), this traditional method is indispensible even today. For cytological observations, however, the method has several disadvantages, among which is the likelihood of cutting a cell into several slices. A long chromosome might be cut into two or three segments, and, even though serial sections can be examined, errors are unavoidable.

Squash preparations eliminate all the defects inherent in sectioned preparations. The intactness of the cell is preserved, allowing no doubt as to the completeness of the chromosome complement or the individual chromosomes. Moreover, the pressure applied during squashing (by thumb or sometimes even by tapping or pounding) forces the cells to flatten, causing all chromosomes to lie on one plane of focus. Frequently, the pressure forces the chromosomes to separate from one another, giving better spreads. In sectioned slides, this was not feasible. If it were not for the squash technique, the polytene chromosomes could not have been used for many studies in cytogenetics and in cell physiology.

The second technical improvement in cytology was the addition of colchicine, an alkaloid from the bulbs of the Mediterranean plant *Colchicum*, to growing tissues prior to fixation. This compound interferes with the formation of the mitotic spindle. When a cell enters mitosis under the influence of colchicine, the spindle fails to form and the cell is therefore "arrested" at metaphase (Blakeslee and Avery, 1937; Levan, 1938; Eigsti and Dustin, 1955). Thus, colchicine treatment increases the number of mitotic cells available for observation of chromosomes.

The question arises as to why mammalian and human cytologists were unable to exploit these useful techniques. Some did, but with poor results. Modern cytogeneticists know that mammalian chromosomes tend to crowd in the metaphase plates, and a hypotonic solution pretreatment (see Chapter 3) disperses the chromosomes to facilitate counting and morphologic characterization. However, if crowding is the only problem, then it should not be difficult to count the chromosomes of species with low diploid numbers. However, as Figure 2.1 shows, a low chromosome number does not ensure an accurate count in mammalian cells. All these figures were taken from squash preparations stained with the traditional acetic orcein. The cells were obtained from cultures of a male Indian muntjac or barking deer, which has a diploid number of $\female = 6$, $\male = 7$, the lowest known of all mammals. Figures 2.1a–c were taken from a preparation without hypotonic solution pretreatment. Figure 2.1a shows a metaphase whose chromosomes appear to be glued together. Such metaphases are extremely common, and there is no way to count them, much less characterize them morphologically. The chromosomes in Figure 2.1b are somewhat more separated, but it is still not possible to count them. The cell in Figure 2.1c, a rare occurrence, displays seven chromosomes if one knows that the cell should contain seven chromosomes. Even so, the morphology of individual chromosomes is extremely poor. In contrast to these figures, the metaphase presented in Figure 2.1d, taken from a squash preparation with a hypotonic solution pretreatment, exhibits seven discrete chromosomes with well-defined morphol-

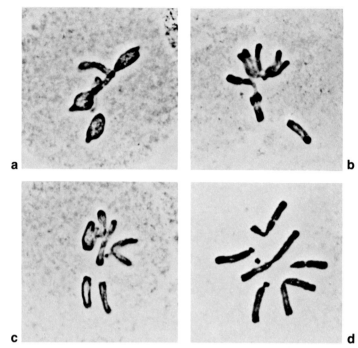

Figure 2.1. Metaphases of a male Indian muntjac (2n = 7). **a–c,** acetic orcein squash preparation without a hypotonic solution pretreatment. Note poor chromosome morphology. **d,** same material with a hypotonic solution treatment before fixation. Note distinct number and discrete chromosome morphology.

ogy, including the secondary constriction regions. The amorphous material associated with the chromosomes without hypotonic solution was found to be ribosome-like particles (Hsu *et al.*, 1965), apparently representing the remnants of nucleoli that disintegrate during mitosis. Hypotonic solution pretreatment, therefore, disperses not only the chromosomes but also the nucleolar material that attaches to the chromosomes, so that a crisp outline can be discerned for each element.

If seven chromosomes are nearly impossible to count, imagine what kind of difficulty early investigators must have encountered when the diploid numbers were in the thirties, forties, and fifties! Even though some of the counts were correct, the morphological delineation of the chromosomes, as can be seen from many camera lucida drawings, tells little about the size, shape, and other pertinent characteristics of the chromosomes. Apparently, the nucleolar material coating the chromosomes in irregular clumps played a detrimental role.

Another difficulty encountered in the studies of mammalian and human chromosomes in the early days was the lack of access to good material. In plants, meristems always contain a good number of mitoses. In insects, ganglia and spermatogonia provide a plentiful supply of division figures. But for human study, it appeared that the only logical material for a reasonable number of division figures were the testes. There were only two ways to obtain human testicular material: waiting outside the operating rooms and waiting by the gallows. Some cytologists tried other materials. Kemp (1929) and later Andres and Navashin (1936) did a considerable amount of work on human chromosomes with tissue-cultured cells, and Chrustschoff and Berlin (1935) tried blood cultures. La Cour (1944) attempted to study human chromosomes with bone marrow smears, but somehow these gallant deviations from the classic sectioned preparations of testicular tissue had little impact on mammalian and human cytogenetics of that period.

It is easy for a modern cytogeneticist to disdain the poor quality of the preparations of the early days, but one must also appreciate the frustrations encountered by these pioneers of yesteryear and admire their spirit. Cytologists could easily walk to greener pastures; the methodology for analyzing plant and insect chromosomes was well established. At the same time, we are disappointed, when browsing through the literature of the early days, to find that most of the efforts on mammalian cytogenetics were really wasted. In the synopsis of animal chromosome studies compiled by Makino (1951), one finds great discrepancies in chromosome counts of the same species, and, after comparing these counts with data obtained by modern methods, practically no useful information. It is somewhat irritating to find that the cytologists of that era were apparently satisfied with their work, since so few of them sought technical improvements.

Perhaps a brief account of the studies of human chromosomes will suffice to illustrate the development of the field of mammalian cytology of the early period. Human chromosomes were observed during the late nineteenth century, but most of the investigations on chromosomes related to the behavior of chromosomes rather than their number, morphology, and relation to cell lineage. In the 1890s, with the emergence of the chromosome theory of heredity, cytologists began to pay attention to chromosome number. However, from the 1890s to the 1920s, reports on human chromosome number varied from 8 to over 50, but the most frequent counts were 24 for the diploid cells and 12 for the haploid. Hans von Winiwarter was one of the outstanding cytologists of this era. He used live tissues for immediate fixation and was the first one to insist on the side-by-side pairing of chromosomes in meiosis. He reported 47 chromosomes in human spermatogonia and 48 in oögonia (Winiwarter, 1912). Thus, he concluded that in man the sex determination mechanism was XX/XO. The discrepancy between Winiwarter's count of 48/47 and the prevailing belief that it was 24 caused much speculation, but the work of T. S. Painter (Figure 2.2) dispelled many doubts that the diploid number of man was in the forties rather than in the twenties. At first, Painter (1922) was not certain whether the diploid number of man was 48 or 46. He seemed to favor 46 because "in the clearest equatorial plates so far studied only 46 chromosomes have been found." He revised his opinion later from 46 to 48 (Painter, 1923), probably to keep Winiwarter's company, but insisted that

Figure 2.2. T. S. Painter.

the sex determination system of man was XX/XY instead of XX/X0 as Winiwarter had proposed.

No one doubted that the human diploid number was 47 or 48 after the 1920s. The sex chromosome argument also slowly died down, as only Winiwarter and his student, K. Oguma, reported the count of 47. All subsequent papers in the 1930s by other cytologists confirmed Painter's interpretation.

I did not have any opportunity to meet these earlier cytologists except Painter. When I came from China to enroll in the Graduate School of the University of Texas, Professor Painter was president of the University and had administrative duties. I did meet him a few times at the Zoology Department annual picnics, which he always attended: But I was working on *Drosophila* then, so our brief conversations usually centered around the salivary gland chromosomes. Painter, of course, was one of the pioneers in this area of research. At any rate, the human diploid number was taken to be 48 until 1956, when the correct diploid number of man, 46, was found. Before then (see Chapter 4) everyone (including myself) was looking for 48 chromosomes, even forcing the count.

After Painter's death, my friend, Margery Shaw, borrowed from Mrs. Painter some slides of human testicular sections prepared by Painter. On several spots there were dried oil patches, apparently the areas observed by Painter. Margery took some photomicrographs and kindly permitted me to reproduce one in Figure 2.3a. This prompted me to remark, in a letter to Malcolm Kottler: "I did get a chance once to see a slide prepared by Dr. Painter, and I failed to make any sense of the twisted, crowded, stacked chromosomes. It's amazing that he even came close!" Figure 2.3b is a reproduction of one of Painter's drawings. It is evident that photomicrography is useless for illustrations of these early preparations, so camera lucida drawing had to be performed to show the cytologists' interpretations.

Because of the technical difficulties in studying mammalian and human chromosomes, very few cytologists dared to enter this dark alley of research. Besides Painter, only two persons, Robert Matthey (Figure 2.4) of the University of Lausanne, Switzerland, and Sajiro Makino of Hokkaido University, Japan, took the endeavor seriously. Both are naturalists by heart, so that their research inclined heavily toward natural history. They have devoted their entire professional lives to

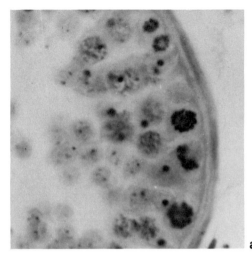

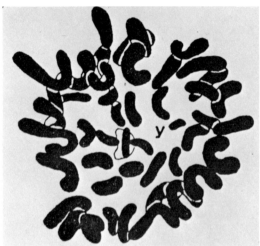

Figure 2.3. a, a section of a human testis with spermatogonial mitoses prepared by T. S. Painter; **b,** a camera lucida drawing of a human spermatogonial metaphase made by Painter.

studies of the chromosomes of vertebrates (and to some extent invertebrates as well). Makino has also been interested in the chromosome constitution of tumor cells.

As Matthey told me, he was born a morphologist. By the age of 6, he was a passionate sketcher of animals and envisioned a future career as a painter or a zoologist. In 1924, he went to Geneva to work in the laboratory of Professor E. Guyénot, whose exceptionally brilliant teaching attracted many students. Guyénot broke with the traditionally systematic review of zoology by teaching general biology—genetics, cytology, and em-

Figure 2.4. Robert Matthey.

brvology—and he incorporated the works of contemporary biologists such as Morgan and Sturtevant. Matthey first undertook eye-grafting of amphibia and then the determination of secondary sex characteristics in reptiles. He also collaborated with Guyénot in studies on regeneration in reptiles. However, Matthey was more attracted by the work of A. Naville, of the same institute, who was working on the evolutionary aspects of sporozoa. It was Matthey's introduction to chromosome studies. He was particularly impressed at the work of K. Nakamura, a Japanese zoologist, who used a mixture of osmic acid and chromic acid as fixative without the addition of acetic acid, which was then considered necessary for fixing nuclear structures. Matthey advantageously used this method until Makino advocated the use of hypotonic solution pretreatment for squash preparations (see Chapter 3.)

Matthey returned to Lausanne in 1929 and remained in the University until his retirement. Although he had meager funding and lacked personnel, he was given a light teaching schedule and administrative workload. In his 40 years at Lausanne, Matthey made every preparation himself, observed every slide himself, and typed every manuscript himself. (I must say that someone surpassed him in the last item. When I was a graduate student at the University of Texas, I wrote a draft of a paper totaling 60-odd pages. I gave it to my professor, J. T. Patterson, to edit, and he complained that my typing was too messy. I told him that I would type a clean copy after he made corrections. For the next couple of weeks, Dr. Patterson was busy with his typewriter, using one finger to type, as always. Then he told me he had finished typing my manuscript.)

In 1949, Matthey published his book *Les chromosomes de vertebres,* which summarized the state of the field up to that time. It contained numerous plates with mitotic and meiotic figures, all drawn by himself. This book is still one of the treasures in my collection.

Matthey made one of his most interesting discoveries in his studies on the chromosomes of the African pygmy mice *Leggada.* Teamed with F. Petter, a mammalogist of the National Museum of Natural History in Paris, he collected more than 300 specimens and analyzed them cytologically. The highest diploid number of the pygmy mouse is 36, all single-armed (acrocentric) chromosomes. However, contrary to the prevailing concept that the chromosome number should be constant within a species, Matthey found an extensive polymorphism in chromosome number, particularly in the populations of the Central African Republic. The diploid numbers of *Leggada* varied from 18 to 36, with all the possible numbers in between, but concomitant with the reduction of chromosome number, there was always an increase in biarmed chromosomes. (This phenomenon was well known as the Robertsonian fusion, namely, two acrocentric chromosomes may fuse at their centromeric ends to form one biarmed chromosome. If each acrocentric chromosome or each arm of a biarmed chromosome is considered one unit, there is no actual change in the number of chromosome arms. A karyotype with 18 biarmed chromosomes has the same number of units as a karyotype with 36 acrocentrics.) Matthey therefore proposed the term *nombre fundamentale* (NF), which appeared in the first issue of the new journal *Experientia* (Matthey, 1945). Cytogeneticists have been using the term ever since.

The best data on mammalian and human chromosomes of the prehypotonic era, painstakingly collected, could not compete with what one can obtain in a few minutes from

a slide casually prepared by modern methods. Even
Matthey and Makino cannot deny that the quality of
earlier works cannot be compared to that of today. But
considering the difficulties (technical as well as financial)
the pioneers encountered, I must salute the knights of the
Dark Age for their courage and dedication.

3 The Hypotonic Miracle

Most mammalian and human cytogeneticists consider the introduction of the hypotonic solution pretreatment for chromosome preparations to be the turning point in vertebrate cytogenetics. Kottler (1974) called it the technical revolution. As many biological historians have commented, the fact that a hypotonic solution can spread metaphase chromosomes was observed by several investigators at least a couple of decades prior to its rediscovery (cf. Kottler, 1974). There were several reasons for the burial of this important discovery:

1. The discoverers, such as Eleanor Slifer (1934), were not chromosome cytologists. They found the chromosome spreading effect of hypotonic solution as a by-product of other investigations.
2. They did not exploit the finding and publicize it by testing it on a variety of materials and publishing papers specifically dealing with the virtues of the technique.
3. Many scientists like to see techniques handed to them on a silver platter. Mammalian cytogeneticists were very complacent about their miserable preparations, as can be witnessed by their repeated claims of accuracy in counts of human chromosomes, which were either 47 or 48. Thus, a phenomenon noted by a grasshopper embryologist did not strike them as a possible way to improve their own work.

Some two decades after Slifer's observations, at least three independent papers appeared in the literature in the same year, reporting the rediscovery of the effect of hypotonicity on chromosomes (Hughes, 1952; Makino and Nishimura, 1952; Hsu, 1952; Hsu and Pomerat, 1953). Arthur Hughes was analyzing the effects of tonicity on chick embryonic cells *in vitro* and found (like Slifer, as a by-product of his research), that hypotonicity spreads the chromosomes of metaphase plates. Hughes, likewise, did not capitalize on this significant discovery.

My own discovery was purely serendipitous (Hsu, 1961). The fact that I even touched on the subject of tissue culture of human cells was a fortuitious event. In 1951, I was scheduled to receive my Ph.D. degree from the University of Texas. As a student of entomology and *Drosophila* genetics, I was looking for a postdoctoral training position in the field of genetics. Unfortunately, opportunities for employment or postdoctoral fellowships were extremely scarce at that time. Several of my professors, particularly Wilson S. Stone, sent feelers out for me but drew blanks. Finally, Professor M. J. D. White told me that there was a possibility for a postdoctoral fellowship at the University of Texas Medical Branch in Galveston, where Professor Charles M. Pomerat had wanted someone to work on the nuclear phenomena of human and mammalian cells in culture. I had no previous contact with tissue culture or cells of higher organisms at all, but out of desperation, I told White that I was willing to give it a try. At the least, so I thought, I would have one year with paychecks coming. White made arrangements for me to visit Pomerat, who fascinated me with his personal charm as well as his time-lapse motion pictures of cell activities, particularly mitosis. This eventful visit changed my entire career.

White was a little uneasy about my leaving *Drosophila* research. Frankly, so was I. For about six months in Pomerat's laboratory, I learned how to set up tissue cultures, take phase-contrast photomicrographs and time-lapse films, and learned a lot of new terminology. But I accomplished very little research. I tried to look at the chromosomes, but found them as crowded and as hopeless as those in histological sections. I liked that laboratory (located in the Texas State Psychiatric Hospital, which we all thought was appropriate), but I was nostalgic, yearning for a return to *Drosophila*.

Then came a miracle. The laboratory had received a few samples of fetal tissues from therapeutic abortions. We set up all sorts of tissues for culture. Many of the cultures were used for time-lapse films, and many were for other experiments, which individual investigators set up on their own. I really did not have definite experiments to run, but I set up some skin and spleen cultures anyway. Then I fixed some of the cultures and stained them with hematoxylin. I really was not looking for anything in particular, but I thought I might be able to see lymphopoiesis *in vitro*. I could not believe my eyes when I saw some beautifully scattered chromosomes in these cells. I did not tell anyone, took a walk around the building, went to the coffee shop, and then returned to the lab. The beautiful chromosomes in those splenic cultures were still there; I knew they were real. I settled down to examine more slides and found that the entire set had unbelievably pretty mitotic chromosomes (Figure 3.1). But none of the mitotic figures appeared to have spindle orientation. Not a single mitotic cell exhibited a side view of metaphase. There were no typical metaphases.

I tried to study those slides, and set up some more cultures to repeat the miracle. But nothing happened. The mitotic figures resumed their normal miserable appearance. I repeated two more times without success. I began to think that there was something "wrong" with

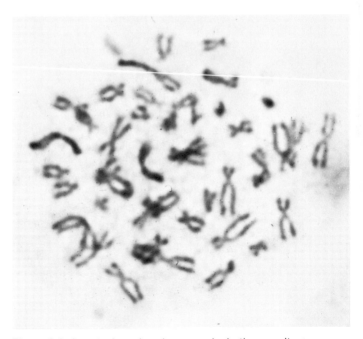

Figure 3.1. A metaphase in a human splenic tissue culture accidentally treated with a hypotonic solution before fixation (Hsu, 1952).

that particular set of human splenic cultures which made the mitotic cells so marvelous. For about three months I attempted to alter every factor I could think of: medium components, culture conditions, incubation temperature, addition of colchicine, fixation procedure, staining method, etc., one variable at a time. But nothing worked until April of 1952, when I changed the tonicity of the balanced salt solution, which everyone in the lab used to rinse cultures before fixation. The miracle reappeared when I mixed the balanced salt solution with distilled water to reduce the tonicity. It was certainly a wonderful feeling to be able to solve a mystery. I realized that I had a powerful tool in my hands and set out to test this tool against materials other than human cells. I found it applicable to all species and all cultures.

It was certain, then, that the first set of human spleen cultures which gave the magnificent mitotic figures had been accidently washed in a hypotonic solution before fixation. The only logical explanation was that one of Pomerat's technicians who prepared the balanced salt solution had misread the scale and prepared a hypotonic solution without knowing it. Since nearly four months had elapsed, there was no way to trace who actually had prepared that particular bottle. Even today, I would love to give a peck on the cheek to that young lady who made an important contribution to cytogenetics, but no one admitted committing an error. Therefore, this heroine must remain anonymous. However, the story I heard years later, that the girl was subsequently fired because she committed an error, is untrue. In fact both Pomerat and I had a difficult time expressing our exhilaration. Pomerat quickly filed a letter of recommendation to the U.S. Immigration and Naturalization Service in support of my application for permanent residence and raised my stipend. I decided that there was a rather fertile field for me to explore with this tissue culture system, and did not touch the subject of *Drosophila* cytogenetics again until 20 years later.

At this juncture, I must mention the dean of the Medical School when I was there, Dr. Chauncey D. Leake, a remarkable man in many respects and one of the few persons whom I revered without reservation. Leake was a scholar and an extremely warm person, yet he was also an imaginative administrator. His office was also in the Psychiatric Hospital on the same floor as our laboratory. It was a relatively large room, but its Spartan interior was unmatched by any executive office I have seen. There was no carpet on the floor and no decorations on the wall. Other than a conference table, necessary file cabinets, etc., there were only two desks—one for him and the other for his secretary, who sat near the door. The door was always open. Anyone could walk in to see the chief without an appointment.

Dr. Leake always kept an interest in the research of Pomerat's laboratory. He regularly attended our seminars and offered comments as a pharmacologist as well as a philosopher. One morning on my way to work I passed by his office. He was reclining in his swivel chair with his feet on the desk, and a cigar in his mouth, gazing at the ceiling. I walked in and asked his secretary: "What's he doing?"

"Shhh. He is thinking. Don't disturb him today," she suggested. So I went on to the lab. I saw him almost in the same pose when I went out for luncheon. But at about 4 o'clock that afternoon, Leake jumped into our lab with a broad smile, waving a sheet of paper in his hand, and shouted to me "I got it." I had no idea what he had until he showed me that piece of paper. He had spent more than half a day composing a poem describing the hypotonic solution pretreatment for chromosome preparations. It read:

CHANCE AND THE HUMAN CHROMOSOMES

In circled nucleus
the twisted rods lie flat
and silhouette their pregnant shapes
for ease in recognition.
A chance mistake
has pulled away the drapes
of overlapped confusion on
the forms and faces of
the human chromosomes,
so now they may
be named, saluted, while
they stay encased
in their old plasmic homes.
In fantasy we contemplate
the permutations which
there might have been,
had these black rods
but polarized and pulled apart.
We think they may,

fermenting quite unseen,
direct and guide
the symphony of life,
which throbs eternally
in every gene.

Leake wrote six poems about tissue culture cells, which he finally collected under the title *Tissue Culture Cadences.* I quote only one here because the rest of them are not related to my subject.

Poems not withstanding, the first thing to do was to take a closer look at the human chromosomes. We "knew" that the diploid number of man was 48, but I thought that the superior morphology in my slides could help me describe them in more detail. This preconception of 48 as the diploid number made the job very difficult, because in many cells I had difficulty in getting the count to equal 48. In order to "force" a 48 count, I decided that some biarmed chromosomes with stretched centromeric regions must be two separate acrocentric chromosomes. There is no need for me to dwell on an extensive analysis of how this error was committed, since Kottler (1974) discusses the subject very well. Briefly, in addition to this most important preconception, there were other technical and psychological factors that caused my unfortunate mistake:

1. The preparations were pretreated with a hypotonic solution alone. The cells expanded like balloons. The metaphase chromosomes were scattered inside the cell, but they were not forced to lie on one focal plane. Therefore, the chromosomes were in a three-dimensional arrangement and in many cases one must focus up and down to characterize a chromosome, resulting in observational errors.

2. The standard tissue culture of that time was the old-fashioned plasma clot method. The cells were embedded in the clot, which made applying the squash technique to flatten the metaphase plate almost impossible. This point was not mentioned by Kottler.

3. Painter was the president of the University of Texas, where I graduated and worked. He was one of the geneticists I greatly respected. His contribution to *Drosophila* cytogenetics had an immense impact, and I benefited from it in my pregraduate and graduate years. It was unthinkable that Painter could be wrong.

When Malcolm Kottler was doing his research on the history of human karyology, he asked me a number of questions. I mentioned to him that not finding the correct human chromosome number was a sore point in my career. Kindly, he consoled me by writing:

> Not that my opinion matters, but I don't think you should consider your "confirmation" of the old result, $2n = 48$, a sore point in your career. Though both tissue culture and hypotonic pretreatment were old, even if rarely used, methods in animal cytology, it seems clear that your reintroduction of both techniques in the early 1950's was a turning point of critical significance.

In my letter to answer some more of his questions, I commented on that point:

> I am not belittling my contribution. I do believe the method I introduced opened the door to the booming progress of the present-day human cytogenetics and mammalian genetics. But after returning an interception for 40 yards and fumbling the ball at the 3-yard line, one always laments that he could have made a touchdown. I sincerely thank you for giving credit to that futile 40-yard run, for it was still the play that saved the game!

It would be unfair not to give credit to Sajiro Makino's discovery of the water pretreatment technique for the improvement of cytological preparations, for he had started this work several years earlier, although our papers were published in the same year. The paper by Makino and Nishimura never received as much recognition as it deserved, probably because they were using tissues *in situ*, so that the chromosome dispersing effect was not as dramatic as those obtained from cells *in vitro*.

Makino (Figure 3.2) went to Hokkaido from Chiba, Japan, with an ambition to learn chicken breeding to improve the breeds for his country. He was advised by an old livestock breeder that knowledge of genetics was very important for the improvement of any breed, so Makino ought to obtain a sound foundation in this field. The livestock breeder introduced Makino to Professor Kan Oguma of the Laboratory of Zoology, Hokkaido University at Sapporo. This incident destined Makino to a lifelong career in animal cytogenetics.

During and after World War II, Japan was beset with financial difficulties at all levels. The most important cytological fixative of that period, osmic acid, was not

Figure 3.2. Sajiro Makino.

available. Makino was forced to find other fixatives and
other techniques, such as the squash technique, as re-
placements, but he found the preparations unsatisfactory. In the summer of 1948 he found that water could
improve cytological preparations (again a fortuitious
event). He was preparing some insect testicular specimens for fixation when the farmers from the villages
came to sell food. Makino had to buy some food since the
farmers did not come every day and no foodstores were
in existence in Sapporo at that time. So he went to the
market and left the testicular specimens in a wet water
basin. He returned to the laboratory a half an hour later
after shopping, nonchalantly took these testicular specimens, and prepared the slides. He was very surprised to
find beautiful metaphase plates with well-preserved
chromosomes forming rosette-like arrangements. Each
chromosome was well defined without overlap or clump-
ing. This incident prompted him to treat specimens with
water deliberately prior to fixation. He tested nearly 50
species of animals, including vertebrates, in the next two
years and, in the majority of cases, found a vast improvement in the quality of his preparations.
    Subsequently, Makino and his disciples found that

hypotonic solutions were better than water because the former are less harmful to the cells. Although it is true that most contemporary mammalian and human cytogeneticists use cell cultures as their material, Makino's method is still the best way to obtain good squash preparations from intact tissues.

In 1953, Dr. Makino visited our laboratory, and it was a real pleasure to receive him. For several years, he had been working on the chromosome changes in the cell populations of the Yoshida sarcoma of the Norway rat. He found a large metacentric marker chromosome in the tumor cells. The rat karyotype was considered to consist of 42 telocentric chromosomes. I set up some rat embryo tissues for culture and found that many chromosomes were biarmed. Since Makino was touring the United States at that time, I invited him to stop in Galveston to see us. He was very excited to see the rat chromosomes with such clarity. I gave him all my slides, and a joint paper resulted from the collaboration (Makino and Hsu, 1954).

Makino suffers considerably from his difficulty in speaking foreign languages. He mentioned this in the preface to his book (Makino, 1975). To compound this difficulty, he also has a hearing problem, an injury, according to him, inflicted by a Japanese army officer during World War II. Being so polite, he usually answered "yes" to whatever question he was asked, without understanding the question. Occasionally, perhaps 5% of the time, he would answer "no." One evening, I took Makino to my apartment with a friend of mine to have dinner, and I did the cooking. As we finished the dinner, Makino stood up, bowed, and thanked me. My friend then asked,

"Well, Dr. Makino, are you now convinced that Dr. Hsu is a good cook?"

"No," he replied and solemnly sat down.

I am sure he was telling the truth, but that is not the usual habit of the polite Japanese people. I simply got hit by that 5%.

In my own case, I was not actively engaged in chromosome research after the first two years in Pomerat's laboratory. Being in a medical school, I was influenced by the constant bombardment of discussions on medical problems, particularly cancer. Pomerat's laboratory also had strings of visitors from all over the world, and their presence also steered me away from research problems relating to cytogenetics. I subsequently found myself working on antitumor serum and tumor transplanatation with Daniel G. Miller and screening cancer chemotherapeutic agents. Moving to Houston to set up my own laboratory in the M. D. Anderson Hospital also slowed me down for a couple of years. In fact, I did very little serious cytogenetic work for several years until Albert Levan came to my laboratory to spend six months with me in 1959. Certainly, I did not regret venturing into the fields of transplantation and chemotherapy because I learned a lot, but I believed that an instant cancer cure was remote, and in order for me to understand cancer I would have to study the genetics of somatic cells, including chromosome structure and function. So I was glad to return to my first love, the chromosomes. But by this time a new era of human cytogenetics had arrived.

4 From 48 to 46

Albert Levan (Figure 4.1) was an accomplished plant
cytologist before he touched animal cells. He was one of
the earliest investigators to study the effects of colchicine
on mitosis, and, in the 1940s, he used onion root tips to
analyze extensively the actions of chemicals and radiation on chromosomes. However, he had been interested
in the chromosomes of cancer cells for some time, because he was impressed many times by the similarity of
the chromosome damage induced in the onion root tips to
the pictures of chromosome disturbances in cancer cells
reported in the literature. Since it was not then feasible to
analyze mammalian chromosomes in detail he did not
venture into this field.

In the late 1940s, Theodore S. Hauschka, then working
in the Institute for Cancer Research in Philadelphia,
transplanted a large number of mouse tumors, and
George Klein converted a number of solid tumors into the
ascites form. The ascites tumor cells remain suspended in
the ascitic fluid, making excellent materials for many
studies. In 1951, Levan took a leave from the University
of Lund, Sweden, to work for a period in Philadelphia.
Hauschka suggested to him that it might be interesting to
take a look at the chromosomes of the ascites tumors. So
they tapped some ascitic fluid of tumor-bearing mice and
made some squash preparations. Levan found good
enough mitotic figures on his slides to warrant continuing
this study. This was the beginning of Levan's contact
with mammalian chromosomes and, I believe, he has
been working in this area of research ever since. The
report (Levan and Hauschka, 1952) gave clear evidence

Figure 4.1. Albert Levan.

for the first time that mammalian chromosomes can be not only accurately counted but also characterized.

In the summer of 1955, Levan came to the United States again, this time to John Biesele's laboratory in the Sloan–Kettering Institute in New York. He was trying to obtain some good chromosome preparations of mammalian tissue culture materials. He received cells of a human tumor line, HEp2, initiated by Alice Moore. Attempts to make chromosome preparations without any pretreatment resulted in poor preparations because all chromosomes clumped together. The next day he used a hypotonic solution pretreatment and was rewarded by finding several metaphases "almost analyzable." The following day, a hypotonic treatment was given to cells that had been pretreated for 20 hours with colchicine in the medium, covering a series of concentrations from 0.0005 to 1 μM. Many cells with analyzable chromosomes were found in the concentration range 0.01–0.1 μM (0.004–0.04 μg/ml). More experiments proved that the combination of colchicine and hypotonic solution pretreatments was also favorable in other human materials, such as the Chang Conjunctiva and Liver cell lines, as well as rat cell cultures.

Figure 4.2. Jo Hin Tjio.

On Levan's return to Lund in August of 1955, Joe Hin Tjio (Figure 4.2) also went there to do some research work. Like Levan, Tjio was an accomplished plant cytologist, then working in Zaragoza, Spain. However, he periodically went to Lund; so Tjio and Levan had been collaborators for years. They decided to apply the new techniques to human embryonic cells and obtained some primary cultures from Rune Grubb of the Institute of Medical Microbiology on the same campus. These cultures had been set up for virus research from lung tissues of legally aborted fetuses. But in these cells, Tjio and Levan were unable to find the expected 48 chromosomes. Instead they found 46.

During the fall and winter of 1955, they accumulated abundant evidence to show that the chromosome number of their human material was 46, not 48. Despite the clarity of their preparations and consistency of their counts, however, the shadow of the "established" human diploid number still haunted the minds of these two investigators. They experienced a brief period of anxiety and even disbelief, and decided to publish their observations (Tjio and Levan, 1956) but to restrict the conclusions to the lung tissues of the four fetuses studied. Tjio and

Levan stopped short of stating that the diploid number of man is 46, but did say, ''it is hard to avoid the conclusion that this would be the most natural explanation of our observations.''

Their apprehension soon diminished when two English investigators, Charles Ford and John Hamerton, informed Tjio and Levan that they had found 23, instead of 24, bivalents in the spermatocytes of three patients (Ford and Hamerton, 1956). There was no longer any doubt that the 46 chromosomes found in the lung cultures were the true diploid number of man, not an exceptional phenomenon.

Furthermore, a member of their own laboratory in Lund, Eva Hansen-Melander, told Tjio and Levan that during the preceding fall she and her husband, Yogue Melander, collaborating with Stig Kullander, had studied mitosis and the chromosomes of liver cells of a human male fetus, and had counted the same number of chromosomes. Using small pieces of liver minced in 3 : 1 alcohol–acetic acid and squashed in 2% orcein dissolved in 60% acetic acid, they repeatedly counted 46 chromosomes, but did not consider the slides good enough to warrant a publication at that time. This independent observation added further credence to the conclusion that the diploid number of man should be revised.

The revision of the diploid number of man was the result of technical improvements. Briefly, Tjio and Levan had used a combination of four techniques:

1. Cell Culture

In culture, the cells do not form a tissue, which would interfere with the effectiveness of treatments like hypotonic solution and slide preparation. The cultured cells can be suspended in the fixative and the staining solution, thus giving an even distribution of a monolayer of cells on the slides and providing an excellent material for the application of an even pressure when squashing.

2. Colchicine Pretreatment

This method was borrowed from plant cytology. Colchicine does two things: It arrests mitosis to increase the number of metaphases, and it allows the metaphase chromosomes to further condense. The chance of excessive chromosome overlapping, which is often a source of error in chromosome counting and characterization, was thus reduced.

3. Hypotonic Solution Pretreatment.

I discussed this technique in the previous chapter.

4. Squashing

Again, this technique was borrowed from plant cytology. As mentioned, the cells in a hypotonic milieu expand like ballons, and the chromosomes are scattered within the cells in a three-dimensional arrangement. Squashing flattens the cells to a pancake-like morphology, forcing the chromosomes to lie in one plane of focus. This not only expedites counting and other characterizations but also drastically reduces errors. Furthermore, the metaphase plates of flattened cells are aesthetically pleasing.

Of these four basic techniques, all but squashing are still used extensively in most cytogenetic work on vertebrates, although investigators may occasionally delete the use of colchicine or its derivative, Colcemid. Squashing has been largely replaced by the air-drying technique (see Chapter 5), but the principle of flattening the metaphase chromosomes into one plane of focus remains unchanged.

The revision of the human diploid number from 48 to 46 came as a surprise to all cytologists and geneticists. But Tjio and Levan's evidence seemed so strong that no one argued with their clarity and accuracy. As Levan once commented "even a child can count 46 chromosomes" in their photomicrographs (Figure 4.3).

Tjio and Levan's startling finding was quickly confirmed by a number of cytologists, including myself, once the "spell of 48" was gone. From these observations, it seemed that the human diploid number was 46 among all cells of an individual, and among all individuals, regardless of race. This consistency was important because some cytologists of that period believed that the chromosome number might be variable. Timonen (1950) had examined squash preparations of human endometrium and claimed that the chromosome number was variable from cell to cell. This report was followed by another paper (Therman and Timonen, 1951) dealing with human embryonic cells. Again, they found a remarkable inconsistency in chromosome number. Even established cytologists such as C. L. Huskins (personal communication in 1953) considered it possible that local tissues such as endometrium might have a variable chromosome

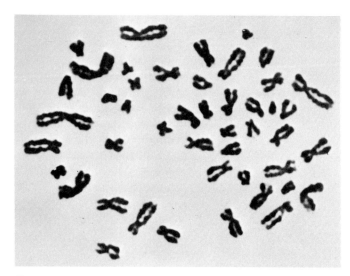

Figure 4.3. A metaphase from a human embryonic cell culture (Tjio and Levan, 1956).

number, but it was too much for him to swallow the notion that embryonic tissues would behave likewise, because it shook the whole foundation of genetics. It was comforting, therefore, to find data on human chromosomes in which constancy, instead of variability, prevailed, and the variability in chromosome numbers apparently could be attributed to imperfections in the techniques (cells ruptured by squashing) and inaccurate counts.

5 The Foundation of Somatic Cell Genetics

In addition to the discovery of the correct human diploid number, which set the stage for the development of medical genetics, several independent research activities during this period (1952–1959) contributed, directly or indirectly, to the field of mammalian and human cytogenetics. These research achievements are not really closely related, but I present them here together for the sake of brevity. I use the chapter title "Foundation of Somatic Cell Genetics" merely from the lack of a more appropriate heading. It is true, however, that all these advancements, especially the method of cloning cells *in vitro* introduced by T. T. Puck, eventually became important tools for the study of the genetics of somatic cells.

Karyology of Neoplasms and Long-Term Cell Cultures

The Boveri hypothesis (1914), that a change in chromosome constitution is the prerequisite for cancer, remained untested because it was not technically feasible to make a critical appraisal of the chromosome constitution of normal tissues. Pathologists and cytologists repeatedly observed numerous mitotic anomalies in cancer tissues and surmised that the chromosome constitution of these cells must be grossly abnormal, but they could not make more definitive statements because they were unable to count chromosomes accurately, much less characterize them in detail.

In the early 1950s, however, some advance in this direction was made by cytologists who used ascites tumor cells. Three groups of investigators, Sajiro Makino in Japan, T. S. Hauschka and Albert Levan in the United States, and Klaus Bayreuther in Germany, worked independently on the chromosomes of rat or mouse tumors. Makino and his associates used the Yoshida ascites sarcoma of the rat to analyze the chromosome constitutions of the tumor cell populations. These investigators found a small proportion of cells containing a large metacentric chromosome, which does not exist in the normal karyotype of the Norway rat. Cells with this marker chromosome steadily increased in frequency through transplantation generations until all cells had this marker. It was unthinkable that the same abnormality could be repeatedly generated *de novo*. The entire cell population of the Yoshida ascites tumor, therefore, must have been derived from a single cell, the stemline.

In the United States, Levan and Hauschka, working on mouse ascites tumors, arrived at the same conclusion, based on their observation that individual tumors possessed a characteristic set of chromosome anomalies (number as well as morphology). In the great majority of cases, the chromosome number was not euploid.

Although the data appeared to support the Boveri hypothesis, one could easily argue that the abnormal chromosome constitutions in long-term transplanted tumors do not necessarily reflect the chromosome constitutions of the original tumor cells. The Yoshida sarcoma was an excellent example. The abnormal marker chromosome was found at first in a low percentage of the cells. Since the tumor was in existence without this chromosome change in every cell, it is unreasonable to use this example to infer that changes in chromosome constitution are the primary cause of malignancy. Of course, neither Makino nor Hauschka and Levan advocated a causal effect, but neither did they maintain that the chromosome constitution of primary tumors must be normal. It was Klaus Bayreuther (1960) who claimed that he found normal chromosome constitutions in numerous primary tumors of the mouse, rat, cow, and chicken. Therefore he opposed, with considerable heat and animosity, the Boveri hypothesis and anyone who supported Boveri. He was so convinced of the correctness of his stand that he took criticisms personally. I kept an open mind on the subject (and still do), but I was more inclined to think that Boveri had a point. If cancer represents genetic changes in a somatic cell, then there is no reason for a mandatory chromosome change to achieve malignancy. On the other hand, chromosome changes certainly can cause profound genetic changes; therefore, it is highly probable that chromosome alterations meet the requirements for neoplastic transformation. In view of the extensive chromosome changes in cancer tissues, it is tempting to think that some, if not all, cancer cells started with a change in their chromosome constitution. I certainly agree with others that it was physically impossible for Bayreuther to analyze *carefully* so many mitotic figures for chromosome constitution, even if he had worked 24 hours a day for two solid years. Thus, his analysis was probably rather perfunctory and overly expeditious.

In the meantime, chromosome analyses of long-term tissue culture cell lines also began. Two of the cell lines, the mouse L strain established by Wilton Earle in 1940 and the human carcinoma line, HeLa, established by George Gey in 1951, were the main targets of investigation. Both were found to be highly aneuploid with a stemline chromosome number between diploidy and tetraploidy, and the chromosome number and morphology were extremely variable. All the normal mouse chromosomes are telocentric, but in L cells, a variable number of chromosomes were biarmed. The abnormalities in number and morphology were reminiscent of the chromosome constitutions of cancers. Of course the HeLa line was derived from a cancer tissue, so the variability was not surprising. But L cells, and other mouse cell lines established by Katherine Sanford, were all derived from normal tissues, and yet the chromosomes of these cell lines were highly abnormal. Thus, it appeared that cells might lose their original euploid constitution, become aneuploid in the course of *in vitro* cultivation, and become malignant (transplantable).

As tissue culture slowly gained popularity, many laboratories carried L cells and HeLa cells as standard material. Without knowing that these cells were extremely hardy and that a small number of them, or even a single cell, could be accidentally introduced to another culture flask containing cells of different origin, many laboratories found that their cell cultures changed characteristics and became established cell lines. Actually, the intruders settled in the new vessel and grew, eventually replacing the original culture. This cellular contamination phenomenon caused a lot of confusion, arguments, and misinterpretations until full-scale detective work was conducted by Klaus Rothfels and collaborators (Rothfels *et al.*, 1959), who used both cytological and immunological methods to prove that the altered cell lines were actually cell contaminations.

This contamination served to warn investigators that tighter control of daily operations in the culture rooms was necessary. In fact, commercial medium suppliers might have been the culprits in some instances, since they used HeLa cells for tests. Laboratories thus might have gotten HeLa cells through the ready-made media they purchased. Another possibility for "cellular contamination" was sheer carelessness by the investigators or technical personnel. For example, wrong labels could have been placed on the culture vessels.

Karyological and immunological characteristics are useful for detecting interspecific cellular contamination and replacement, such as HeLa cells contaminating a rabbit cell culture or L cells contaminating a human cell culture. It was more difficult to detect intraspecies contamination, since chromosomal and antigenic features were the same. The problem was partially resolved years later by Stanley Gartler (1967), who used isozyme patterns to identify cellular contamination. Not unnaturally, such announcements disturbed those whose precious cell lines turned out to be HeLa. But the evidence was so strong that cell contamination in cultures is an undeniable fact. The more laboratories employ cell cultures for experiments, the more such cases occur. Some cytogeneticists even devote full time to checking cultures for other laboratories.

Cloning of Cells *in vitro*

Some years ago, a physicist friend told me that he missed the golden age of physics of the 1930s because he was too young, but he would not miss the golden age of biology. I am afraid that he also missed it. It is a mystery why some physicists and physical chemists venturing into biology make sparkling contributions and others do not. It is not because of differences in intelligence. Perhaps there is a difference in approach. Physicists entering biology should at least think somewhat like biologists in order to tackle biological problems. Certainly, Theodore T. Puck behaves both as a physicist and a biologist. Originally educated as a physical chemist, Puck had considerable experience in engineering before his first taste of biology in Max Delbrück's laboratory. After he took a position at the University of Colorado Medical School, he began to think that perhaps mammalian cells could be handled *in vitro* just like microorganisms. One of the key advantages of using microorganisms for genetic studies is their ability to produce single-cell colonies or clones. Puck reasoned that if mammalian cells *in vitro* can be handled like bacteria and produce clones, there would be hope for using such cells for genetic analyses. It is true that Katherine Sanford had isolated clones of strain L mouse cell line in the 1940s, but her method was too laborious to obtain quantitative data. Puck's introduction (Puck *et al.*, 1956) of a plating method provided a quantitative approach to the assessment of mutation rates, survival rates under adverse environments, and many other conditions.

Cells *in vitro* can survive and proliferate if a relatively large number of them are present initially to help one another in their efforts to condition the medium or if a single cell is confined in a tiny space with a very small amount of medium. Sanford's micropipette method for cloning followed the latter principle. Puck's original method was to use the so-called feeder layer. When the cells are X-irradiated at a very high dose, they are still capable of performing normal metabolism, but their chromosomes are so shattered and their mitoses so abnormal that ordinary cytokinesis cannot take place. All the cells develop into giants and finally degenerate. However, this process takes many days, and in the meantime, the cells condition the medium. If a small number of nonirradiated cells are added to the culture vessel containing the irradiated (feeder) cells, these healthy cells are able to develop into individual cell colonies because the feeder layer feeds the healthy cells yet the feeder cells are themselves reproductively sterile. The feeder layer was later found to be unnecessary for certain long-term cell lines such as HeLa, but in difficult cases, it is still very useful.

Naturally much technical improvement has been made since Puck's first experiment, but the principle remains the same. As a direct application of the cloning experiments to chromosome research, Ernest H. Y. Chu (Chu and Giles, 1958) analyzed several clones of the HeLa line isolated by Puck and found that most clones had a very narrow range of stemline chromosome numbers, which indicated homogeneity, whereas the parental HeLa cells had a wide spectrum of chromosome numbers. One of the clones, however, had a distribution curve indicating heterogeneity as great as that of the parent, suggesting that karyological (hence genetic) stability may vary from one cell to another.

Technical   Probably Alma Howard and S. R. Pelc (1953) never an-
Improvements ticipated that their crude experiment on autoradiography
with Vicia bean roots would expand into a new field of
science, cell kinetics, with myriad applications, including
cancer therapy. In chromosome research, autoradiography had its heyday in the 1960s after J. Herbert Taylor
used tritiated thymidine to analyze DNA replication
phenomena (Taylor et al., 1957). Some of the pertinent
data will be presented later.

Other than autoradiography, several minor technical
improvements were made in chromosome cytology. One
was a study of the effects of hypotonicity on chromosomes by David Hungerford (Hungerford and DiBerar-
dino, 1958), who found that hypotonic KCl had some
cytologically desirable effects. Most contemporary
human cytogeneticists use KCl to treat lymphocyte cultures for chromosome preparations. It appears that KCl
at 0.075 M is ideal for expanding the cells, and the sister
chromatids of each metaphase chromosome usually lie
close to each other.

There were other improvements related to slide preparations. As previously mentioned, the squash technique
had been in existence for many years. Most plant
cytologists and insect cytologists were adroit squashers.
One problem with squash preparations, however, was
that they were temporary and slowly deteriorated. Therefore, the dry ice method for making squash preparations
permanent (Conger and Fairchild, 1953) was a welcome
improvement. I guess most readers pay little attention to
fine print, but the paper did give credit to Jack Schultz for
inventing the technique. Alex Fabergé told me that he
had learned the procedure from Schultz in Philadelphia
but Jack never published the procedure. When Alex visited the Oak Ridge National Laboratory, he told Alan
Conger about the technique.

To squash mammalian cells after a hypotonic solution
pretreatment requires some adjustments. Plant tissue,
with its tough cellulose cell wall, needs a strong force on
the coverslip to crush and flatten the cells. Mammalian
cells are much softer and tapping or pressing too hard
breaks cells. Timenen encountered this problem in his
work on human endometrium. The cells were broken and
the mitotic chromosomes were scattered over the slide in
small groups.

Therefore, the pressure must be started gently and
increased slowly as the excess staining solution (usually

acetic orcein) is squeezed out and blotted off. Some persons whose thumbs cannot bend upward have trouble doing squashes. Hans Stich is one of them. When Hans was working in my laboratory as a guest investigator, I actually saw him laying all slides on the floor and stepping on each with his heel, muttering "I must get my wife down with me the next time I come." Indeed he did, and I found Kikki to be one of the world's best squashers, in addition to being a beautiful woman.

The pain of not being able to make good squash preparations with mammalian cells was substantially relieved by the air-drying technique. Again, Alex Fabergé told me that he had been using the air-drying preparations for pollentube mitosis for quite some time. The trouble with Alex is that he seldom describes his findings. Thus, cytogeneticists waited until Klaus Rothfels and Louis Siminovitch (1958) discovered the procedure in the course of their study on monkey chromosomes. They grew the rhesus monkey cells on regular microscope slides so that they could fix the cells directly after pretreatments. Somehow they neglected to take care of the slides after fixation and the fixative dried out. Subsequently, they found the chromosomes well spread and in one plane of focus. This air-drying technique is now so popular among mammalian and human cytogeneticists that squashing is almost a lost art, although, in my laboratory, we still use squashing for certain experiments.

6 Funny Looking Kids

The literature of genetics repeatedly describes individual plants or insects with an extra chromosome (trisomic) or with one less chromosome (monosomic). In *Drosophila melanogaster,* for example, individuals with three X chromosomes have been known about for decades. Similarly, in the jimson weed *Datura stramonium,* trisomics were known to affect many characteristics of the plant, and trisomy for each chromosome was linked to a consistent series of morphological abnormalities. In the tobacco plant, monosomics have been used by geneticists to locate genes in a particular linkage group.

But the finding of counterparts in man should indeed be considered a monumental discovery, because the etiology of many congenital syndromes, with their multiple but characteristic defects, had baffled clinicians. In fact, if it were not for Jérôme Lejeune's discovery of a trisomic condition associated with mongolism, human cytogenetics would probably have died soon after the correct diploid number was determined and the chromosomes characterized. Lejeune's report (Lejeune *et al.,* 1959) created a new field of medicine—medical cytogenetics.

Speculations that some congenital anomalies in man such as mongolism might be caused by chromosomal abnormalities were made as early as the 1930s. Said Waardenburg (1932): "I would like to urge cytologists to consider the possibility that in mongolism lies an example of specific chromosome aberration in man . . . perhaps a chromosome deficiency through nondisjunction or the reverse, a chromosome duplication." Bleyer (1934) also suggested trisomy as the possible cause of this common developmental defect. Ursula Mittwoch (1952) actually examined spermatocytes of a mongol patient but failed to find aneuploidy.

Lejeune (Figure 6.1) had been studying mongoloid children for some time before embarking on a cytological study. Noting that the overall topological picture of these patients is reminiscent of lower primates, he considered that a chromosomal error might explain the sudden change of a polygenic response. But his intention to check the chromosomes of his patients began only after he heard Jo Hin Tjio's report at Copenhagen that human chromosomes could be counted and characterized with accuracy and clarity and that the diploid number of man is 46.

Nowadays, many young biologists think that lots of equipment and supplies are prerequisites for research. They probably would not even think about starting their research without an ultracentrifuge, a scintillation counter, an electron microscope, and access to a computer. These are indeed wonderful to have, but look at what Jérôme had when he started his historic project in

Figure 6.1. Jérôme Lejeune.

1958. He needed cell cultures for cultivating his patient's tissues *in vitro*; naturally, he did not have the facilities. Fortunately, Mlle le Dr. Marthe Gautier, one of Professor Raymond Turpin's assistants, knew the techniques and was able to help. Lejeune's laboratory had no running water, but he could use a tiny kitchen-like room next door. He had no microscope, so he asked the Bacteriology Laboratory to donate a discarded microscope, whose gears were so worn that its arm could not be adjusted. (Lejeune resorted to using a piece of tinfoil from a chocolate candy wrapping to stabilize the gears.) He then arranged with the Pathology Department to use their photographic equipment once a week for two hours. From the start to the first publication of his finding of trisomy, he spent a total of a few thousand francs, or about $200.

Since the lymphocyte culture method was not available in 1958, Lejeune had to ask the patients' parents for permission to use biopsy tissues for fibroblast cultures. He explained to the parents that the examination would be absolutely useless for their children, but might increase understanding of the cause of the disease. All the parents cooperated very willingly.

In that year, Jérôme got the first good look at the chromosomes of a 2-year-old boy with mongolism, who still visits the clinic twice a year. The chromosome number was 47, trisomic for one of the smallest elements now known as No. 21. Lejeune had expected to see a deletion or a monosomy. Thus, because preconception did not enter the picture, the observation must have been correct. A greater surprise came when an identical chromosome picture was obtained from the second and the third patients.

In September of 1958, Lejeune left Paris for Montreal to participate in the International Genetics Congress. He had all the slides and pictures with him but said nothing at the Congress. After the meeting, Clark Frazer invited Lejeune to give a seminar at McGill University, where he presented his data. As far as Jérôme could tell, Frazer was deeply interested in this report, but he could not read the reaction of other people in the audience. Lejeune surmised that they probably thought he was raising a lot of noise for no good cause.

Even after his paper was published, Lejeune still could not believe that all mongols had the same trisomy until confirmation from abroad arrived. It was a relief when he exclaimed *Mais, c'est vrai!* Now the conclusion is firmly established: Mongolism is associated with trisomy of one of the smallest autosomes, designated as No. 21.

Some people may wonder why I, a Chinese, keep using the somewhat racist term mongol or mongolism instead of the more neutral term Down's Syndrome. It is unfortunate that an ignorant racist term was coined to begin with, but unless one proposes a new term, changing mongolism to Down's Syndrome does not eradicate ill feelings, because it was Down who invented the term "mongolian idiot." One may as well use the original term, so long as he or she realizes the folly. This reasoning also explains my reluctance to adopt the current move of changing words to eliminate "sexist" language. I believe terms are less important than what is in the mind. To male chauvinists, changing "chairman" to "chairperson" does not erase their prejudice. Perhaps the story I heard about mongolism would make Orientals less resentful: a Westerner once asked a Japanese pediatrician how he would describe the features of mongols. Said the doctor, "Look like Caucasians."

Lejeune's report hit the scientific community like a storm. People not only confirmed his findings but extended them to other congenital disorders, particularly those with multiple anomalies. The "funny looking" kids became the prime targets of inquiry; but inmates of mental institutions and criminal institutions also contributed heavily to our knowledge of human cytogenetics. Because of their association with abnormalities with sex chromatin, several kinds of congenital maladies received early attention. As a result, the relationships between the abnormalities in sex chromatin and sex chromosomes were clarified (see Chapter 8). There are too many persons to whom credit should go; I have therefore only cited some of the more pertinent papers in the course of the story. The reader may consult many excellent reviews and comprehensive books (German, 1970; Hamerton, 1971; Makino, 1975) for details. I would, however, like to present a brief account of one person, W. M. Court Brown, for his contribution to human genetics. Court Brown was probably the least known among human cytogeneticists of that period. In technical papers, his name seldom appeared as senior author; he let his associates receive the glamour and placed himself in the background. It was Court Brown who created the Medical Research Council Unit in the Western General Hospital of Edinburgh, Scotland, where many important advances have been made. Investigations by Baikie, Bucton, Harden, Jacobs, McLean, Tough, and others were all, in large measure, made possible by the background and atmosphere created by Court Brown.

His early career was fairly orthodox. After graduating in medicine from the University of St. Andrews, he specialized in radiotherapy. He decided, however, to abandon a successful career in medicine to devote himself to full-time research. He joined the Medical Research Council in London and rapidly gained recognition for his epidemiological approach to the long-term harmful effects of ionizing radiation. In 1956, he moved to Edinburgh to set up his own research team. At that time, he made two decisions which determined the future development of his work. First, he believed that only by the cooperation of clinician and basic scientist could real progress be made in medical research. Second, he believed it was necessary to begin by examining the genetic material of the cell in order to understand the fundamental mechanisms underlying the induction of tumors.

I was told by David Harnden that Michael Court Brown was a complicated person. He would not have wished for any posthumous platitudes, and it would be senseless to pretend that he was a nice, friendly guy. He was, however, one of the most interesting, unusual, and stimulating individuals. His forthright honesty often led to troubled personal relationships, but it also earned for him a reputation for outstanding scientific integrity.

I have discussed Court Brown because I think that the Edinburgh Unit played a significant part in the development of human cytogenetics, and it was Court Brown's leadership that bore the fruit. Although Court Brown died at the peak of his career and many of his old colleagues left Edinburgh, the laboratory under the new leadership of H. J. Evans has another excellent group of investigators who have already made many contributions to modern human genetics, cytogenetics, and molecular cytogenetics.

Of course, Edinburgh was not the only laboratory that made significant contributions to the development of human cytogenetics in the early 1960s. Many investigators in Europe and in the United States discovered numerous interesting (and perhaps also alarming) facts. The cumulative data showed that human chromosomes are highly unstable and that practically all types of chromosome aberrations found by cytogeneticists in plants and insects occur spontaneously in human populations with a frightening frequency. I shall give only a very brief account of the types of abnormalities found in man:

1. Trisomy

Other than Trisomy 21 (mongolism) and sex chromosome anomalies (see Chapter 8), there is trisomy 13 and trisomy 18, discovered by Klaus Pätau and his associates (Pätau *et al.*, 1960; Smith *et al.*, 1960) in the United States and John Edwards *et al.* (1960) in England. These two trisomies produce such characteristic and striking morphologic features that it is no longer necessary to check the chromosomes for diagnostic purposes.

2. Monosomy

The only monosomy known to date is Turner's Syndrome (X0, Chapter 8). Apparently, autosomal monosomies are lethal.

3. Fusions

As mentioned previously (see Chapter 2), fusion is the extreme case of chromosome translocation in which two single-armed (acrocentric) chromosomes translocate together in toto with very little loss of genetic material. This fusion can cause two acrocentric chromosomes to become one biarmed chromosome. In the human karyotype, there are 10 acrocentric chromosomes, and fusion between any two of them is possible. The resulting karyotype contains 45 chromosomes instead of 46, but the carrier of such an abnormality remains viable and fertile. However, in meiosis of these individuals, disjunctional abnormalities may become more frequent, giving rise to aneuploid gametes. Some parents of mongoloid children were found to have fusions.

4. Reciprocal Translocations

When a translocation involves segments of chromosomes instead of total chromosomes, the genetic material of the carrier is again preserved and the individual is normal. However, such individuals may produce abnormal gametes which, when fertilized with a normal gamete of the opposite sex, would produce genetic imbalance in the offspring. The offspring may be deficient in one segment of a chromosome or may have a duplicated segment carried by an unrelated chromosome. However, these arrangements are not always detectable by conventional staining methods. Many of these were unequivocally identified after the banding techniques became available, but numerous obvious cases were discovered prior to banding.

5. Deficiencies

Although reciprocal translocations can cause a deficiency in one chromosome and duplication in another, deletions of a chromosome segment can occur without translocation. The most well-known case for deletion in man is the *Cri du chat* syndrome discovered by Lejeune and associates (Lejeune *et al.*, 1963). In addition to other morphological abnormalities, the babies of this syndrome cry like a kitten. A segment of the short arm of chromosome 5 is invariably missing in these children.

6. Polyploidy

The first case of triploidy was discovered by Böök and Santesson (1960) in a Swedish boy. The child had two types of cells, those with 46 chromosomes (XY) and those with 69 chromosomes (XXX). Probably because he was a mosaic diploidy–triploidy, he survived after birth.

7. Embryonic Wastage

Many chromosomal abnormalities cause such genetic imbalance that the zygotes seldom survive beyond the early periods of gestation. Thus, in spontaneous abortions, particularly in early stages of pregnancy, a high incidence of chromosomal abnormalities would be expected. Indeed, the work of David Carr in Canada (1967) and the Boués in France (Boué and Boué, 1966) provide ample evidence to confirm the suspicion that lethal chromosome errors of various kinds, including tetraploidy, occur at a high frequency in abortuses.

The explosive activity in human cytogenetic research revealed that the frequency of chromosome aberrations in human populations is much higher than biologists had expected. The collective efforts of cytogeneticists and clinicians have made great strides in the search for the etiologies of many perplexing congenital defects. However, I would like to echo the sentiments of Jérôme Lejeune who, in a recent letter, says, "I cannot refrain from some feelings of melancholy. So much is known about these chromosomes, and so little has been achieved to help these cheerful children! The hope of my life is to live long enough to see one of them cured of mental deficiency and become a human geneticist!"

7 Of Beans, Weeds, and Human Cytogenetics

Cell culture proved to be a superior system for karyological investigations, but it required considerable experience and a sizable expenditure. It was not practical or even feasible for many laboratories to set up cell culture facilities just for studying chromosomes. Furthermore, it is extremely difficult to procure biopsy specimens from normal persons, except in surgery cases. Therefore, the lymphocyte culture method discovered by Peter Nowell (1960), as a byproduct of another research project, was one of the most timely and welcomed contributions to human cytogenetics. Only a small quantity of peripheral blood sufficed to yield a good number of mitoses for chromosome analysis. Peter, a pathologist teaching at the University of Pennsylvania, was studying leukocytes in culture. After several days of incubation of the cultures he obtained an unexpected result: a large number of mitotic cells. Leukocytes in the peripheral blood were regarded by most biologists as terminal points of cellular differentiation. Thus, they were not supposed to possess the ability to undergo mitosis. In Nowell's cultures, however, many cells looked like lymphoblasts instead of mature lymphocytes. He thought that some factor in the tissue culture system caused this anomalous behavior. Therefore, he systematically tested the variables that he could think of, but initially failed to find the responsible factor. It turned out that tissue culture per se was not responsible for stimulating the lymphocytes to return to a more immature form. Nowell was using the popular method for separating erythrocytes and leukocytes prior to cell culture (a method used by Edwin

Osgood for many years), agglutinating erythrocytes with phytohemagglutinin (PHA), a crude powder extracted from the common navy bean, *Phaseolus vulgaris.* Phytohemagglutinin not only agglutinates red cells to facilitate separation but also, it turned out, stimulates mature lymphocytes to return to a blastic state capable of cell division once again. When Nowell used other methods to separate red cells from leukocytes, no mitosis could be observed in his cultures.

After Pete submitted his manuscript to *Cancer Research,* he received the reviewers' comments. One of them said, as Pete recalls, "It is an interesting observation but of no conceivable significance to science." Ironically, a compilation made by *Current Contents* on the most widely cited papers in cancer research during the past 20 years, ending 1974, lists Nowell's paper (1960) as number one. It not only gave cytogenetic laboratories an easy and inexpensive method for obtaining ample mitoses for chromosome analysis, but also made prospective donors much less reluctant to provide samples since few persons would object to giving a small quantity of blood. Furthermore, it was a clear demonstration that highly differentiated cells are capable of returning to a less differentiated state if proper environmental stimuli are present. Many biologists took advantage of this phenomenon to analyze the biochemical events that take place in the dedifferentiation process of lymphocytes *in vitro.*

Nowell's discovery realized one of J. B. S. Haldane's dreams. Haldane (1932) wrote: "A technique for the counting of human chromosomes without involving the death of the person concerned is greatly to be desired. It seems possible that satisfactory mitoses might be observed in a culture of leukocytes." Indeed, such a technique quickly became available as Moorhead *et al.* (1960) described the steps employed for human cytogenetic studies. After over 15 years, the method is basically unchanged.

For a period of time, the demand for phytohemagglutinin was so high that the suppliers depleted their stock. Some laboratories (including mine) prepared their own bean extracts for leukocyte culture. Since this mitogen aspect of cell physiology was new, manufacturers monitored the PHA preparations by determining their agglutinating ability. This was not reliable because agglutination and mitogenic effects depend on two different substances in the bean extract. However, this complication was soon removed and more purified PHA samples to be used exclusively as lymphocyte mitogen became available. Some manufacturers even made kits containing everything required for blood culture and slide preparation so that all one needed to do was draw blood and incubate the cells for three to four days. This allowed institutions without a tissue culture laboratory, including high school biology classes, to obtain satisfactory human chromosome preparations.

The beans are not the only plants that contain a substance capable of stimulating mature lymphocytes to return to their blastic state. In the mid-1960s, another mitogen was found by Patricia Farnes and Barbara Barker, of the Rhode Island Hospital at Providence, Rhode Island. Their paper (Farnes *et al.*, 1964) did not mention how this pokeweed mitogen (PWM) was discovered. Pat Farnes kindly supplied me with background information, which I think is worth recording.

The incident occurred in 1961, when a 3-year-old girl was admitted to the Rhode Island Hospital with mysterious ailments. She had been admitted to another hospital for several days with vomiting and diarrhea, after which she developed central nervous system disturbances. Finally, she suffered respiratory failure and died.

After the postmortem examination, the neuropathologist brought Pat and Barbara a slide of brain autopsy material to their attention. In this preparation they found "lymphocytes" that looked blastogenic. They then asked the pediatricians about the case and were informed that the girl had all the early symptoms of pokeberry poisoning. Upon questioning the parents, the pediatricians found that the little girl had a habit of popping things into her mouth and swallowing the objects without chewing. There was a large pokeweed plant (*Phytolacca americana*) in her family's yard and apparently the girl swallowed a considerable number of the berries. Pokeberries are not palatable, so children usually chew one or two and stop. This particular child apparently swallowed quite a quantity, because when she first vomited, many berries and seeds were noted.

Armed with this information, Pat thought that maybe pokeweed has a mitogen. So she and Barbara picked a load of the berries and put them in the freezer. Somehow they did not get around to testing the mitogenic effect of pokeweed berries for two years. They considered the chemical work to be too laborious for a research project that might not yield any result. Another reason for their hesitation was the neuropathologist's insistance that the child died of viral encephalitis. These factors understandably dampened their enthusiasm.

As Pat stated,

If it had not been for Lydia Brownhill, who had come to Rhode Island Hospital from Kurt Benirschke's group at Dartmouth, we probably would never have given the pokeberries another thought. One day we were talking to her about PHA and she mentioned grinding the beans and using the supernatant.

This reminded them of the frozen pokeberries. It appeared to them that throwing some juice into a blood culture to take a look at the lymphocytes was not a difficult and expensive project. Therefore, they smashed the berries, put the supernatant through a filter, and dropped the juice into some buffy coat cultures. In three days, there were blasts in all cultures. Although the investigators were convinced that the pokeberries contained a mitogen, the medical community steadfastly refused to accept the connection between ingestion of the berries and the death of the child. Pat says that the neuropathologist (now deceased and recalcitrant to the end) said that the pokeweed had nothing to do with the case. The parents, who must have had guilty feelings about the whole affair, told everyone that the child died of a virus infection, not pokeberry poison. Other people at the hospital shrugged and said "so what? PHA is a good mitogen; why do we need another one?" Such comments were somewhat myopic and premature. Many cytogenetic laboratories began to use PWM because it does not produce so much agglutination in the cultures as does PHA. Some laboratories also use PWM as an alternate to PHA for blood samples that respond poorly to the latter.

For cytogenetics, the lymphocyte mitogens are now indispensable fixtures. Nevertheless, leukocyte cultures are not panaceas. They have some disadvantages. For example, they could not be maintained as long-term cell lines to provide a continuous supply of the same cells. It was only during recent years that establishing long-term lymphoid lines became feasible (see Chapter 15). Also, it is undesirable to obtain a huge quantity of blood from a single person to perform large-scale experiments. Furthermore, lymphocytes of many animals do not always respond to stimulation by these mitogens. More work directed toward establishing cell lines that can be propagated for many generations *in vitro* is still needed.

From electrophoretic mobility experiments, Farnes and Barker found that PWM and PHA were two different substances. Perhaps those who shrugged off the discovery of the PWM a decade ago should shrug no more, because recent studies indicate that PHA stimulates only T cells, whereas PWM stimulates both T and B cells. At a certain concentration level, PWM stimulates only B cells. Sophisticated investigations are now in progress in many directions, relating to problems of differentiation, membrane chemistry, immunology, and numerous other subjects. The more kinds of mitogens we find, the more we can understand these important processes. And all these advances stemmed from some innocent observations in some cytogenetic laboratories.

8 Sex and the Single Chromosome

In the late 1800s, microscopists began noting condensed chromatin pieces embedded in a generally diffused chromatin matrix of interphase nuclei. These condensed chromatin bodies have been known as the chromocenters. In the late 1920s and early 1930s, Emil Heitz made a series of careful cytological observations on a number of organisms and concluded that there are two major types of chromatin: euchromatin, which condenses during cell division and decondenses (becoming diffused) during interphase, and heterochromatin, which remains condensed all the time. The chromocenters are equivalent to the heterochromatin. However, it was difficult to identify the chromosomal locations of heterochromatin because at metaphase, when the chromosome morphology is the best, both euchromatin and heterochromatin are fully condensed. In interphase, when differentiation is best, recognition of individual chromosomes is not feasible. Heitz determined the locations of heterochromatin by using prometaphasic chromosomes. Here, the chromosomes are morphologically identifiable yet the euchromatin is not fully condensed. In *Drosophila melanogaster,* heterochromatin is located in the proximal third of the X chromosome, the entire Y chromosome and the centromeric areas of the autosomes.

Extensive genetic data revealed that heterochromatin contains very few, if any, structural genes. Thus, both cytological and genetic evidence appeared to suggest that heterochromatin is *intrinsically* different from euchromatin. However, this conclusion did not seem to be universally applicable. For example, the X chromosome of the grasshopper is strongly condensed in the male meiotic prophase but does not show such behavior in somatic cells of male or female individuals. It appears that the X chromosome of the grasshopper can be either heterochromatic or euchromatic, depending on the tissue. In this case, then, the heterochromatic appearance of the X chromosome in male meiotic prophase may be a physiological response, not intrinsically different from other chromosomes.

Perhaps the chromosome behavior of the mealy bugs studied by Spencer Brown and his disciples conclusively established that heterochromatic appearance does not a priori indicate a lack of genes in a chromosome or a segment of a chromosome. In the mealy bug *Planococcus,* an entire haploid set of chromosomes was found to assume a heterochromatic appearance and to be genetically inactive in the male. In the female individuals, however, both sets showed normal behavior. Physiologically, therefore, the male mealy bug is really a haploid with only one functional set of chromosomes, the maternal set. Yet this functional set, when transmitted to the next generation, would become functionless as well as heterochromatic in the male offspring again. Heterochromatin, therefore, appears to be of two basic types— one containing structural genes but cytologically condensed and physiologically inactivated (the mealy bug type, now known as the *facultative heterochromatin*) and the other containing no structural genes (the original heterochromatin, now known as the *constitutive heterochromatin*).

All these appeared to be exceptional systems in special groups of organisms and not applicable to human and mammalian genetics. But an observation made by Murray L. Barr (Figure 8.1) back in 1949 (Barr and Bertram, 1949) proved that facultative heterochromatin is not limited to grasshoppers and mealy bugs. It occurs in human tissues as well. The discovery was entirely unexpected. But observant investigators can usually capitalize on unexpected findings to make greater contributions than they expected for their original research.

Murray Barr was with the Department of Anatomy of the University of Western Ontario, London, Canada, starting a research program in neurocytology. He noted some early reports that nerve cells displayed structural changes following prolonged activity. The experiments had not been very well controlled, but the results were

Figure 8.1. Murray L. Barr.

nevertheless interesting. For example, in the early part of this century, Hodges and others had compared the appearance of neurons in resting pigeons with those of homing pigeons after a long flight. From 1939 to 1945, Barr served as a medical officer with the Royal Canadian Air Force. His thoughts turned to the homing pigeons when he noted the behavior of fatigued pilots trying to locate home base in foggy England after long night bomber runs over the continent. Although experimental and practical aspects were far removed, the RCAF thought that it could afford $400 for a research project of a recently retired wing commander and gave Murray, who returned to academic life, that amount of money to have equipment made for a proposed experiment. The idea was to stimulate the hypoglossal nerve of cats on one side; the constituent motor neurons would presumably be activated by antidromic, as well as peripheral conduction, and the adjacent resting hypoglossal neurons in the medulla would be close by for comparison.

At this juncture Ewart (Mike) Bertram, who had just received a B.S. degree in biology, applied for graduate work. Barr and Bertram went ahead with the experiments together, stimulating the hypoglossal nerve of cats for eight hours and examining the neurons at various intervals after stimulation. In studying the Nissl-stained sections that first became available, Bertram noted that the stimulated cells showed chromatolysis of the Nissl material. He further noted that a small chromatin mass that was usually next to the nucleolus in the control neurons was very often some distance from the nucleolus in the stimulated neurons.

Barr and Bertram divided the subsequent work between them, Bertram quantitating the degree of chromatolysis and Barr working on nuclear changes. This division of labor was a fateful one, for Murray became heavily engaged in a research field he had not dreamed of. His decision to work on the nucleus was influenced by T. O. Caspersson's demonstration at the International Union Against Cancer, where Caspersson discussed the role of the nucleolus. So Murray wanted to determine what changes the nucleolus underwent during the chromatolytic process. This involved measuring its diameter with a filar micrometer eyepiece in many experimental and control cells at various times following the period of stimulation. Concurrently, he recorded whether the chromatin body was adjacent to the nucleolus, free in the nucleoplasm, or, as sometimes happened in later stages, adjacent to the nuclear membrane.

As the work progressed, Murray found that the small chromatin body, which, for want of a better term, he called the nucleolar satellite (we now call it the sex chromatin or the Barr body), was present in the neurons of some cats but not in others. The presence or absence of this chromatin body could not be correlated with the experimental procedures. There was no explanation for the puzzling discrepancy until, when working in the lab one night, he took a good look at this records and noted that the "satellite" was present in the nerve cells of female cats but not in those of male cats. He and Bertram ran through some more experiments and found that the correlation held up, that is, the chromatin body is found only in female neurons.

This discovery of cytological sex steered Murray's

research career from neurology to cytology for nearly 20 years. He believed that the cytological sex difference must be of fundamental importance and was willing to invest his time and effort in the problem. Consequently, the next few years were spent on a rather tedious program of finding out how universal this cytological sex difference might be. Bertram finished his Master's degree and went to the University of Buffalo for his Ph.D (at that time Western Ontario did not have a Ph.D program), but other graduate students joined Murray to explore various aspects of this research. Briefly, these were the findings:

1. The presence or the absence of the sex chromatin was not restricted to the neurons. Nonnervous tissues, both adult and fetal, possessed the same characteristic.
2. Mammals other than the cat (other carnivores, artiodactyls, and primates, including humans) showed the same sex difference.
3. In rodents, the nuclear sex difference was obscured by other chromatin bodies in the nuclei, and in the Virginia opossum, both sexes were found to posses a piece of chromatin, but it is larger in female cells than in male cells.
4. The sex chromatin was Feulgen positive, indicating that it contained DNA.

The nuclear sexing phenomenon, once firmly established, quickly found medical applications, for example, the determination of the basic sex of many hermaphrodites. To investigate these, Barr and his students and assistants, notably Keith Moore, first developed the skin biopsy procedure, and then the easier buccal smear method. They found that female pseudohermaphrodites were chromatin-positive and male pseudohermaphrodites, chromatin-negative. This was, as Murray puts it "no great discovery, but the paper resulted in literally hundreds of skin biopsy specimens being sent to me from all over the world for interpretation."

Then, in a male with dysgenesis of the testes, or Klinefelter's syndrome, the cells were found to be chromatin positive, similar to female cells. Noticing that some of the men with the syndrome were somewhat mentally retarded, Barr's team surveyed thousands of males and females in institutions for the retarded in Ontario. They found sex-chromatin-positive males as well as females with two and even three sex chromatin bodies.

Although the sex chromatin technique became widely used in the clinics, many basic questions remained unanswered. Conceivably, the anomalies were related to the sex chromosomes, since in mammals the sex determination mechanism was known to be of the XX/XY type. Then what did the sex chromatin represent: an X chromosome, two X chromosomes, or parts of the X chromosomes? If it represented a single X chromosome, did it represent the maternal or the paternal one? The sex chromatin certainly appeared to be a chromocenter (or heterochromatic). Was its genetic activity also inactivated in a way similar to the mealy bug chromosomes? Why was sex chromatin positive in some male hermaphrodites and why in some women was there more than one sex chromatin? Why were women with Turner's syndrome sex-chromatin-negative?

Lejeune's discovery of trisomy in mongols (see Chapter 6) quickly stimulated biologists to investigate patients with sex chromatin anomalies. The Edinburgh group of investigators (notably Patricia Jacobs and David Harnden), the London group of investigators (Charles Ford, Paul Polani, Ursula Mittwoch, etc.), the Swedish investigators (Martin Fraccaro, Jan Lindsten, S. Bergman, etc.), and, later, many other investigators contributed significantly to our understanding of sex chromosome anomalies. A male with Klinefelter's syndrome and positive sex chromatin was found to have 47 chromosomes and an XXY sex chromosome constitution. A sexually underdeveloped female with Turner's syndrome had 45 chromosomes with an X0 sex chromosome constitution. Women with more than one piece of sex chromatin were found to possess multiple X chromosomes. The number of X chromosomes was invariably one higher than the number of sex chromatin pieces, for example, XXX females had two sex chromatin bodies and XXXX females had three sex chromatin bodies. The same formula could be applied to Klinefelter's syndrome, for example XXXY males had two sex chromatin bodies and XXXXY males had three.

From the chromosome data collected from patients with sex chromosome anomalies, it appeared that each sex chromatin represented one X chromosome or a portion of an X chromosome. It also appeared that one X chromosome (one of the two in normal women or the only one in normal men and in women with Turner's syndrome) did not behave this way. Almost simultaneously, genetic data on the mouse collected by Mary Lyon (1961) and Liane Russell (1961) strongly suggested that in female somatic tissues, the inheritance pattern of sex-linked genes resembled that of heterochromatin in *Drosophila*. Lyon proposed a hypothesis that one of the X chromosomes in female tissues is inactivated and that the inactivation may occur either in the paternal or the maternal X early in embryonic development.

This fascinating hypothesis quickly stimulated many biologists to test whether the same phenomenon (inactivation of one X chromosome) occurred in human somatic cells. The best example was probably the experiments of Davidson *et al.* (1963), who isolated cell clones from individuals heterozygous for the activity of the sex-linked gene, glucose-6-phosphate dehydrogenase (G6PD). Indeed, the assays showed that some clones showed G6PD activity and others did not. The Lyon hypothesis was now well established: One of the two X chromosomes of female somatic cells, at least in eutharian mammals, has no genetic activity, similar to the situation found in the mealy bugs. Euchromatin may be physiologically turned into heterochromatin (facultative heterochromatin), a process now called "Lyonization" by some biologists. It also explains why a woman carrying four X chromosomes per cell may appear quite normal.

Cytologically, it was still difficult to identify the inactivated (or repressed) X in metaphase figures. It was particularly difficult in the human complement, because the X chromosome has a morphology similar to many autosomes. But a somewhat indirect method—autoradiography using tritiated thymidine as a label—was developed at that time. After Taylor, Woods and Hughes (1957) introduced this compound, many biologists used the autoradio graphic technique to study a variety of problems related to DNA synthesis of higher organisms. In rye, it was known that constitutive heterochromatin is located at the terminal areas of each chromosome. Antonio Lima-de-Faria (1959) discovered that heterochromatin of rye replicates later than euchromatin. Taylor (1960), using Chinese hamster cells in culture, also found that the DNA synthetic sequence appeared to be preprogrammed, with some chromosome segments replicating earlier than others, the sex chromosomes being among the last to replicate. These results suggested that ^{3}H-thymidine autoradiography might be a useful tool to identify heterochromatin of higher animals, and thoughts naturally turned to the facultatively heterochromatic X chromosome. James German (1962) demonstrated that indeed one of the C-group chromosomes invariably replicates later than all others. Morishima *et al.* (1962) reported the same phenomenon. For a number of years hence, thymidine autoradiography was used extensively to identify heterochromatins, both constitutive and facultative. It was not a convenient method, but it was the only one available. Even at the present time, when we have a good procedure to demonstrate constitutive heterochromatin (see Chapter 18), we still use autoradiography to identify the inactivated X chromosome.

Before the discovery of sex chromosome anomalies in man, the sex determination mechanism of *Drosophila* was thought to be applicable to higher animals also. In *Drosophila*, XXY individuals are females, and X0 individuals, males. Although the basic sex determination system of man is also XX/XY, the reverse is found in multiple sex chromosomes (XXY male and X0 female). Even though the human species breeds like flies, their sex determination systems are not exactly the same.

9 The Denver Conference and Beyond

After the correction of the human diploid number by Tjio and Levan, several teams of cytologists began to analyze the human karyotype in detail and to propose nomenclature systems. I was not involved in this type of activity until Albert Levan spent six months in my laboratory in 1959. It was a real pleasure to work with Albert because he is not only a great cytologist but also a wonderful person. With only one functional eye, he is more observant with a microscope than any cytologist I know, and is extremely meticulous in his work. He would never take a photograph to measure the chromosomes from an enlarged print. Rather, he would move each chromosome to the center of the viewfield and measure its camera lucida image to minimize spherical aberration from the optics. At the time of his visit, I happened to have a skin fibroblast line from a normal boy, a son of a mongoloid woman. We therefore used the cells for karyotype analysis and proposed our nomenclature system.

Naturally, every team devised its own system of nomenclature. Thus, confusion arose, and newer investigators did not know which system to follow. Charles Ford, then working in the Medical Research Council Radiological Research Unit at Harwell, England, thought it important to call a conference to formulate a standard nomenclature system. All the arrangements were made by T. T. Puck and his associates at the University of Colorado Medical School. For practical reasons, Ted Puck decided to keep the group as small as possible and to limit the participants to those who had already published human karyotypes. He proceeded to apply for some funds from the American Cancer Society for this meeting and to send out invitations.

Puck realized that scientists might be protective of their own ideas, systems, and interpretations. In cases of disagreement, arguments might unnecessarily prolong the meeting or even go beyond the academic level. Therefore, at the outset he decided to invite some geneticists to serve as counselors. His criterion for these persons was a great reputation but no human chromosome research. The idea was that they would be impartial arbitrators in case of lingering dispute. Puck first considered choosing two such moderators, but decided to choose three because two moderators might also get involved in an unresolvable argument. D. G. Catcheside (Birmingham), H. J. Muller (Bloomington), and Curt Stern (Berkeley) accepted his invitation, with Professor Catcheside serving as chairman.

The conference was held in Denver in April, 1960. The group was relatively small, but international. The participants included, in addition to the three counselors, J. A. Böök (Uppsala), E. H. Y. Chu (Oak Ridge), C. E. Ford (Harwell), M. Fraccaro (Uppsala), D. G. Harnden (Edinburgh, T. C. Hsu (Houston), D. A. Hungerford (Philadelphia), P. A. Jacobs (Edinburgh), J. Lejeune (Paris), A. Levan (Lund), S. Makino (Sapporo), T. T. Puck (Denver), A. Robinson (Denver), and J. H. Tjio (Bethesda).

The meeting lasted four days. Progress was made despite many heated disagreements. It was amazing to witness the emotional involvement over minute details. One of the early tasks was to find a concordance among the six different nomenclature systems already in existence. The next was to devise a way of arranging the chromosomes in some sort of order. Fortunately, human chromosomes fall into certain natural groupings according to size and morphology. Most published systems followed such natural groupings, but different teams of investigators used different ways to arrange the groups.

Most investigators used Arabic numerals to designate the chromosome pairs. Jérôme Lejeune's group, however, used letters for each group and designated numbers within each group, for example, G1, M3. Lejeune advocated using letters, adding that such symbols can be interpreted with various meanings in any language. For example, the symbol G may symbolize grand, great, or something else. Jo Hin Tjio quickly pointed out that it means nothing in Chinese. The participants finally agreed to use numerals for each chromosome pair, and the numbers were to be arranged consecutively from 1 to 22, with the sex chromosomes designated by the symbols X and Y. However, the decision to dispense with group symbols by the Denver Conference was an unfortunate one, since cytologists later found the lettered group symbols to be very useful. In the Denver document, the recommendation was to refer to each group by the numerals of the chromosomes within the group, using the two extremes connected by a hyphen, for example, Group 1–3, Group 6–12, etc. This system created some inconvenience, particularly with Group 6–12, to which the X chromosome also belonged. Therefore, when referring to any chromosome within the group, one must call the group 6–12 +X. However, Lejeune's symbols, G, M, T, P, C, V, etc. were also inconvenient. Klaus Pätau (1961), in commenting on the lettered group symbols, proposed alphabetical designation of the groups. This nomenclature was accepted by many human cytogeneticists.

At any rate, in the Denver document, the classification of the human chromosomes can be summarized as follows (this is the actual Table 1 of the report):

Group 1–3:   Large chromosomes with approximately median centromeres. The three chromosomes are readily distinguished from each other by size and centromere position.

Group 4–5:   Large chromosomes with submedian centromeres. The two chromosomes are difficult to distinquish, but chromosome 4 is slightly longer.

Group 6–12:   Medium-sized chromosomes with submedian centromeres. The X-chromosome resembles the longer chromosomes in this group, especially chromosome 6, from which it is difficult to distinquish. This large group is the one which presents major difficulty in identification of individual chromosomes.

Group 13–15:   Medium-sized chromosomes with nearly terminal centromeres ("acrocentric" chromosomes). Chromosome 13 has a prominent satellite on the short arm. Chromosome 14 has a small satellite on the short arm. No satellite has been detected on chromosome 15.

Group 16–18:   Rather short chromosomes with approximately median (in chromosome 16) or sub-median centromeres.

Group 19–29: Short chromosomes with approximately median centromeres.

Group 21–22: Very short, acrocentric chromosomes. Chromosome 21 has a satellite on its short arm. The Y chromosome is similar to these chromosomes.

A karyotype is presented in Figure 9.1 as an example.

The participants woked hard to reach a sensible, yet flexible, nomenclature system, and proceeded to write a report. The draft was read and reread, corrected and recorrected, and, by the afternoon of the third day, it was complete. The participants felt a sense of relief, as if a historic document was being written. Indeed it was; but thinking in retrospect, I could easily appreciate the difficulty of the American forefathers in arriving at the nation's constitution. Here we were worrying about how to name the 23 pairs of human chromosomes, not the welfare of the country and its people, yet it took three days to reach some agreement.

On the morning of the fourth and last day, the report was read three times and, after some additional changes, everyone signed it. A relaxed atmosphere prevailed, with the participants reporting in turn their new research findings. Personally, I was pleased to have an opportunity to meet a number of human cytogeneticists from Europe for the first time.

The Denver Conference report, entitled "A Proposed

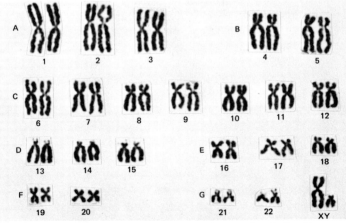

Figure 9.1. A human karyotype arranged according to the Denver nomenclature system.

Standard System of Nomenclature of Human Mitotic Chromosomes," was published in several journals in order to reach the widest range of readers. It urged cytologists to adopt the proposed system for the sake of uniformity. Of course, it recognized that cytologists had the right not to adopt the standard system, but it recommended a reference be made in published works to the standard system.

The influence of the Denver Conference went beyond the immediate environment of its participants and the standard nomenclature system for the human chromosomes. It exemplified cooperation among scientists to solve a problem, set the stage for future conferences as knowledge of cytogenetics accrued, and stimulated cytologists working on chromosomes of mammals other than man to acquire similar systems. Indeed, three subsequent conferences on standardization of human chromosome nomenclature were held, always in conjunction with the International Human Genetics Congress.

The conference following the Denver Conference was held in London in 1963. There were 24 participants, nine of whom were members of the Denver Conference. The London report made very few changes in the nomenclature, but it made one unfortunate and ambiguous description (at least in my view), namely, a "secondary constriction" on the proximal regions of the long arms of chromosome Nos. 1 and 16. Theoretically, the term *secondary constriction* denotes any attenuated area other than the centromere region (primary constriction). However, secondary constriction has been customarily used by cytologists to denote the nucleolus organizer region. As we later found (see Chapter 18), the secondary constriction of chromosomes 1 and 16 are not nucleolus organizers. In the human karyotype, nucleolus organizer regions are located on the short arms of the D and G groups of acrocentric chromosomes, which the London Conference referred to as "satellite stalks." The London report also adopted the alphabetical group symbols proposed by Pätau.

The next standardization conference on human cytogenetics was held in Chicago in 1966. This time, 37 cytogeneticists and four "observers" representing different governmental and private organizations participated. The National Foundation—March of Dimes published the report "Chicago Conference: Standardization in Human Cytogenetics" as one of its Original Article Se-

ries (Chicago Conference, 1960). This report was pref-
aced by the introductory address of Professor L. S. Penrose and included the two previous records, the Denver
and the London Conferences. This book is not the place
to give details of many useful suggestions made at the
Chicago Conference. The readers may want to obtain a
copy of the report for permanent reference.

During the six years between the Denver Conference
and the Chicago Conference, human cytogenetics had
undergone considerable improvement and refinement,
but there were no dramatic changes in terms of removing
ambiguities and increasing resolution of chromosome
recognition. Many pairs within a chromosome group still
could not be differentiated unequivocally. The world had
to wait a few more years for the revolutionary improvement brought about by banding (see Chapters 16–19).

The Paris Conference, held in 1971, made numerous
new recommendations, including recommendations for
the various banding patterns and standardization of the
idiograms, that is, the nomenclature system for each
band (Paris Conference, 1971). In the Paris Conference, a
standing committee was created (chaired by John L.
Hamerton) to take care of various recommendations,
new developments, and many other chores in order to
present a workable agenda for future conferences (Paris
Conference Supplement, 1975).

Without question, many more advances in human and
mammalian cytogenetics and genetics will be made, since
these fields are in an exponentially growing period. But it
was the Denver Conference that set a precedent for a
sensible way to avoid chaos in the literature.

10 The Occasional Lion Tamer

As mentioned earlier, Robert Matthey and Sajiro Makino spent much of their lives studying mammalian cytogenetics and cytotaxonomy in the prehypotonic era. But the application of improved techniques (tissue culture, colchicine, hypotonic solution, etc.) to the studies of mammalian chromosomes really came about after human cytogenetics had been well established. Many human cytogeneticists occasionally worked on the chromosomes of one animal species or another as a hobby, but most of them stopped there. Only a few cytologists made serious attempts to analyze the chromosomes of various mammalian taxa. Michael Bender and Ernest Chu independently studied many primate species. Charles Ford and John Hamerton started some excellent work on the shrew populations of Europe. Others included David Hungerford, who worked on artiodactyls, Charles Nadler on Sciuridae, Alfred Gropp and T. H. Yosida on rodents, and Karl Fredga on carnivores. It is interesting to note that none of these investigators is a bona fide mammalogist. Those mammalogists who finally ventured into cytogenetic research belong to the younger generation.

I believe two laboratories, Kurt Benirschke's and mine, did more ploughing than others in the field of mammalian cytogenetics in the early and mid-1960s. In my own case, what stimulated me to work on mammalian chromosomes were a couple of fortuitous events. I have been a naturalist at heart ever since I was a youngster; therefore these incidents simply rekindled my first love.

Every summer, my institute accepts a number of bright high school and college students to work in various laboratories to get a taste of what research is actually like. I usually took one student each summer. In 1961, an error in communication sent three students, one boy and two girls, to my laboratory. I had no idea how to "entertain" these kids, to make their stay worthwhile, to let them learn something, and at the same time to keep them out of my hair.

The human lymphocyte culture method of Paul Moorhead and co-workers was available at that time. I thought each could try to take one animal species and culture the lymphocytes for chromosome analysis. Thus, they could learn cell culture, microscopy, photography, and other techniques. So I told them to talk to the superintendent of the Houston Zoological Garden to see if they could get blood samples from some of the animals.

They made a telephone call, and were referred to the veterinarian of the zoo, George F. Luquette. Luquette did not know exactly what they wanted, so he called me to find out more details. Apparently, he was interested. I told George (with whom I later had a wonderful association and friendship) that I did not know exactly what I wanted either, but thought we would just try some blood cultures for chromosome studies.

I thought we would start with some gentle creatures, such as antelope and deer. Luquette rejected this idea because these animals are too easily frightened. He suggested we start with the big cats. I was reluctant, but he assured me that there was no danger. So the next day, everyone in the lab went to the zoo, where Luquette already had a lion in a squeeze cage. I thought the tail might be a safer place for blood samples, and one of my technicians was brave enough to try. She poked around for a while but did not get much blood in her syringe. The lion was not happy, of course, and neither was Dr. Luquette. "You amateurs get out of here and let me handle him," he ordered. So we all left the premises and waited outside. Within five minutes, George came out with about 10 ml of lion blood in his syringe and told us that the best place to bleed a lion is from the paw. To this day, I still marvel that he never lost his own paw.

At any rate, we got the blood and proceeded to set up cultures. To our disappointment, we found no mitosis in any of the slides. We went back to the zoo a number of times for more blood samples of lions, tigers, leopards, pumas, and cheetahs, but found only a few mitotic figures in all the slides we prepared.

By that time the two-month stay for the students was up, and I gladly bid goodbye to all of them. But I was frustrated by not having succeeded. The cat business had become an obsession and a challenge; so I talked with Luquette some more, and we initiated the skin biopsy procedure. We finally managed to procure sterile skin biopsies from these animals for cell cultures. For a period, I visited the Houston Zoo two or three times a week to obtain biopsies. Once, an important administrator of the University of Texas came from Austin to meet staff members of our institute, and I was one of the few who could not attend the meeting. My department chairman informed him that I sent my apologies but could not be there because I had an appointment at the cat house.

Perhaps I should made a final note here about George Luquette. George was attending law school when our collaboration was going on and was also working in his spare time at an animal hospital to take care of pets other than dogs and cats, such as eagles, snakes, raccoons, and lions. He once called me from the animal clinic saying that he needed some differential blood counts for a 5-year-old pet lioness. While we were there, I thought it might be fun to have my picture taken with this patient. Her owner kindly used his last sheet of Polaroid film to document that I occasionally tame lions as a hobby (Figure 10.1). Subsequently, the lioness hurt a boy who went too close to her cage to pet her. This time, Luquette

Figure 10.1. T. C. Hsu and patient.

the attorney defended the animal in court and won the case.

I soon discovered that zoos are good places for obtaining biopsy materials from spectacular animals, but for a systematic study, they are unsatisfactory in many respects. Many mammalian species are never kept in the zoos because they do not attract visitors. Among a couple of thousand species of rodents, only a few exotic forms are represented in the Houston Zoo. Out of hundreds of species of bats, the zoo maintains only the vampire bat *Desmodus* from Mexico and the flying fox *Pteropus* from India. In fact, the first bat species we studied cytologically, the Seba's fruit bat, *Carollia perspicillata,* was a discard from the Houston Zoo.

With the methods for procuring tissue biopsies and cell cultures slowly improving, I believed that a survey of the karyotypes of mammals might be a useful endeavor. I had two purposes. One was a naturalist's approach—to compare closely related species or genera to discover the variability in karyological characteristics. Trends of karyological evolution might emerge, and the data might also help mammalogists as an additional set of characteristics for distinguishing species and other taxonomic categories. The second purpose was to attempt to find materials useful for cytological research, such as species with very low diploid numbers, very high diploid numbers, distinguishable chromosome morphology, special features such as unusual distribution of heterochromatin, nucleolus organizers, and sex chromosomes, and perhaps unexpected bonuses.

A friend of mine, Robert E. Stevenson, who was working in the National Institutes of Health in the early 1960s, paid me a visit when he was in Houston on official business. He knew I was looking for cytologically favorable materials and had heard that the Mongolian gerbils had 22 chromosomes, which he suggested I check out. I finally managed to get two specimens and found Bob's information inaccurate. The gerbils have a diploid number of 44 instead of 22.

Apparently, Bob tried to spread the word around for me. Later I received a letter from William Holdenreid, then working in the Animal Resources Branch of NIH, expressing his concern and displeasure over my taking the gerbils as experimental material. He thought I should not breed animals of foreign origin since accidental escape of the animals might upset the local fauna. He further suggested that I undertake a survey of the karyotypes of North American mammals both for my own purpose and for collecting information for fellow scientists. I told him I was willing to do such a survey but knew of no mammalogists who could help me in procuring and identifying specimens. He kindly gave me a list of over a dozen names to whom I could send out proposals. I then wrote each of these people a letter, explaining in detail my project and asking for help. I fully realized that most mammalogists would not be interested in such an adventure, but a few might. Indeed, one letter came back expressing enthusiasm. It was from Murray L. Johnson, of the University of Puget Sound, Tacoma, Washington. Murray is a surgeon by profession and a mammalogist by avocation. He does enough surgery to make a living and spends the rest of his time teaching and performing research on mammals. We finally became collaborators and great friends. He periodically supplied me with biopsy tissues and live animals for my work. Being a surgeon, he took excellent biopsies.

The rest of the persons I contacted politely declined. I later found out that to contact established mammalogists is not an effective method; graduate students are much better targets. They frequently go on field trips, are more enthusiastic, and may be looking for new fields to explore. I finally got a chain reaction started. My old collaborator, Paul Moorhead, organized a summer course to teach cytogenetic techniques and asked me to be on his faculty. He ran two consecutive summers of three weeks each at the Rhode Island Hospital in the mid-1960s. Other faculty members included Bill Mellman and Frank Ruddle. I enjoyed participating in that course, but the most beneficial event to me was an opportunity to meet James Patton, then a graduate student at the University of Arizona. Jim was the only person on the student roster who did not hold a doctorate degree, but he was the only person who gave me much assistance. His major advisor, William Heed, is a *Drosophila* geneticist who overlapped both Paul and me in our own graduate school days under the same major professor, J. T. Patterson. In fact, Bill Heed was Dr. Pat's last student, admitted when Pat was 73. It was because of this relationship that Jim Patton was accepted for the summer course. Jim was interested in exploring cytogenetic problems of mammals rather than *Drosophila,* but had no one to help him at the University of Arizona.

Jim Patton showed me some of his photomicrographs taken from bone marrow preparations of some rodents. They were not impressive. I offered to help him in cytology if he would help me in mammalogy, including procurement of specimens. It turned out that I got the better end of the bargain. After a few trips between Houston and Tucson, I soon became acquainted with many of Jim's fellow graduate students and friends, particularly Robert J. Baker, who was interested in the chromosomes of bats. Later, a snowball effect gave me assistance from many more young mammalogists. It was a pleasure to know their approaches, enthusiasm, and sincerity. Several of them (Patton, Baker, Alfred Gardner, and Dean Stock) spent varying periods of time in my laboratory, and James Mascarello finished his degree with me. Although Amara Markvong is not really a mammalogist, her work in *Mus* chromosomes done in my laboratory in the mid-1970s had a heavy touch in that direction. Amara probably represents my swan song in mammalian phylogeny, since I do not plan any more research in systematics.

I believe we also have achieved our second original goal, to find cytologically favorable materials. I shall defer presenting a summary of phylogenetic studies (Chapter 23) until the banding techniques are discussed, because interpretations based on gross chromosome morphology alone have led to erroneous conclusions. However, I will append here that this innocent project received some undue notoriety. After we finished determining the karyotypes of the specimens, we knew that one day, such materials would be handy for one experiment or another. But it is physically and financially impossible to maintain growing cell cultures in the laboratory, especially with the number of animals we set up daily. Luckily, cryobiology was sufficiently advanced in the 1960s so that cell culture laboratories could store viable cells in a freezer if a certain set of protocols was followed. We used crude CO_2 freezers for many years to store our cultures before we finally were able to order liquid nitrogen freezers. At first, we did not freeze all the specimens because of the lack of freezer space. We regretted this later since we were not able to obtain biopsies again from these animals. But we did store many of them, from aardvark to zebra.

One thought came to our mind: Wouldn't it be wonderful if some prehistoric creatures had stored viable cells of extinct animals (Saber-toothed tigers, for example) and we could revive them in culture to study their chromosomes and DNA. Then our thoughts turned to some of the endangered species of our own time. Perhaps the cells of cheetahs, rhinocerous, fin whales, and polar bears may last a thousand years when such animals no longer exist in the world. Our future generations may still have their viable cells and DNA to play with. Perhaps biological science will be so advanced then that the processes of development and differentiation are completely understood, and technology so perfect that it will be feasible to dedifferentiate cultured fibroblasts into totipotent cells for redifferentiation. All these assumptions are, of course, science fiction at the moment, but the possibility exists. Thus, the "frozen zoo," as *Time* magazine referred to it in an article in 1971, may have some additional uses in the future. But the recent sensationalism of the news media, which calls my collection the "clone zoo," really twists both the meaning and the intention.

11 The Pathologist Who Went Astray

Kurt Benirschke (Figure 11.1) started his mammalian cytogenetic work in an oblique way. His interest in cytogenetics began with the problem of twinning in man. He was a pathologist at the Boston Lying-In Hospital from 1956 to 1960, and was interested in the fact that more twins died than single babies. This led him to a greater interest in the twinning process per se, and he read a considerable amount of literature on this topic, including freemartinism. Since the marmosets and the armadillos were known to be polyembryonic, Benirschke was also interested in these animals. He and his co-workers found that many marmoset twins are indeed chimeras, that is, twin embryos that have exchanged cells. They proved this conclusion by examining the chromosomes of these animals: A certain proportion of cells in females have XY chromosomes and a certain proportion of cells in males have XX chromosomes. Kurt's interest in armadillos later led him to an adventure from Brazil to the Paraguay jungle by illicit means, trying to catch a giant armadillo. He hired a fisherman to row him upstream to cross the border and almost went to prison on his way back. Result: no catch.

While in the Boston clinic, Benirschke was also frequently consulted about infertility, and noted a fair amount of testicular atrophy among many patients. This led him to an interest in the infertility of the mule. In 1960, soon after he moved to Hanover, New Hampshire, to assume the chairmanship of the Pathology Department of Dartmouth Medical College, he was asked to speak to a group of pediatricians in a neighboring community. At

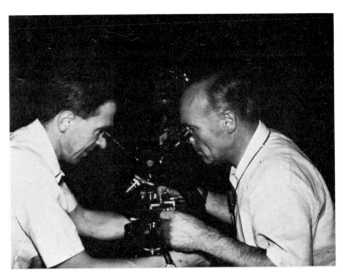

Figure 11.1. Left, Kurt Benirschke; right, Charles E. Ford. (Courtesy of Dr. Frances E. Arrighi.)

the end of his talk, he asked the audience whether anyone knew of some mules from which he could draw some blood. One of the doctors in the audience led Kurt to his farm to procure blood samples of his mules. Recalls Kurt: "I remember distinctly the Christmas of 1960, with my family cutting out chromosomes of the mules."

Benirschke consistently got 63 chromosomes from the mules, but he had no idea of the chromosome constitutions of the parents—the horse and the donkey. So he proceeded to ask owners of these animals for blood samples. He finally finished his study and found that the diploid number of the horse is 64 and that of the donkey, 62. In the meantime, in Susumo Ohno's laboratory, Jose Trujillo did a similar piece of research and obtained similar results. However, Benirschke went on further to investigate the chromosomes of all the living horses and zebras. According to Kurt, this research was not planned. He was invited to give a talk at Cincinnati the next summer, and he took his family with him on that trip, intending to return to Dartmouth by way of Canada. But he discovered later that he had forgotten his passport. Fearing that he might encounter problems returning to the U.S., he persuaded his family to visit the Catskill Game Farm in New York state instead. He had read an article in a magazine about the efforts of the people at the Game Farm in animal conservation. He approached the director, Dr. H. Heck, in his best German and tried to find out whether he could bleed a Przewlaski horse, which is regarded as the most primitive of the living horses. Dr. Heck hesitated, however, and offered an exchange, namely, to study the chromosomal relationships between the South American camels, namely, llamas, gnanacos, vicunas, and alpacas. The deal was that if Kurt would solve these problems, he would be permitted to bleed the Przewalski horse.

Kurt therefore proceeded to carry out his end of the bargain. It was disappointing to him and to Heck that all species showed identical karyotypes, but at least he was now given the opportunity to examine the chromosomes of the precious horse species. The diploid number of the Przewalski horse turned out to be 66 (Benirschke *et al.*, 1965), the highest of the living horses.

It appeared that all the species in the horse family Equiidae had high diploid numbers. However, no information about the chromosomes of the zebras was available at that time. In a magazine Benirschke saw a picture of a zebronchii (a hybrid between a zebra and a donkey) born in Manila. He obtained some tissue samples of that animal for cell culture and found 48 chromosomes. This suggested to him that (*a*) the diploid number of the zebra must be low (should be 34) and (*b*) animals with such dissimilar chromosome constitutions could hybridize. Thus, he went on to analyze the karyotypes of all the zebra species and found that they indeed had low diploid numbers.

Because of the Przewalski horse adventure, Kurt had an easier time in gaining access to the animals kept in the Catskill Farm, and, as a result, he and his colleagues made karyotypic analyses of many species of artiodactyls, including the incredible Indian muntjac (barking deer). Like the horses, most deer species possess high diploid numbers. The muntjacs are a group of relatively small, reddish deer that characteristically utter a noise similar to the barking of a dog. The diploid number of the Reevese (or Chinese) muntjac, *Muntiacus reevesi,* was found to be 46, all telocentric. In fact, this count is one of the lowest among the deer species. But when Kurt and his colleague, Doris Wurster, finally obtained cell cultures of the Indian muntjac (*M. muntjak*), they simply could not believe what they saw: seven chromosomes (cf. Figure 2.1). They kept quiet for two or three years because they thought something was wrong with their tissue culture system or their cytological procedure. But when they obtained a couple more specimens, they confirmed that the diploid number of the Indian muntjac is indeed ♀ = 6, ♂ = 7 (Wurster and Benirschke, 1970). This is one of the most unusual cases in mammalian karyology: closely related species show such unbelievable differences in chromosome constitutions. How the 46 chromosomes of *M. reevesi* changed into six of *M. muntjak* is a mystery (or, as Robert Matthey calls it, a scandal). By centric (Robertsonian) fusions, the largest reduction of the chromosome number that can be achieved from 46 would be 23. Obviously, translocation involving total chromosomes are not limited to the centromeric ends. Many recent reports on human cytogenetics indicate that tandem translocations involving centromere–telomere and telomere–telomere may be more common than cytogeneticists had previously expected. At any rate, this byproduct of research in cytotaxonomy of the ungulates made some contributions to biomedical sciences, because a number of biologists (including myself) now utilize the Indian muntjac cell lines established by Doris and Kurt for a variety of research projects.

While the collaboration with the Catskill Game Farm continued, Benirschke entered the second phase of his work on mammals—the carnivores. It was stimulated by a call from Clint Gray of the Washington Zoo. Dr. Gray was going to immobilize most of the carnivores for vaccination and was willing to take biopsy specimens for chromosome studies. Kurt persuaded Doris to spend one week in Washington procuring the tissue specimens. It is easy to imagine that this short trip led to an enormous amount of work, but of course the eventual dividend was also high. Because of their increased contact with the zoos, they accumulated much information on the karyotypes of many mammals.

I must add here that, as a pathologist, Kurt Benirschke did not completely go astray. At Dartmouth and later the University of California at San Diego, he continued his activities as a pathologist as well as a cytogeneticist. His interest in the placenta pathology is well known. For a period of two or three years, however, he did devote all his time and effort to mammalian cytogenetics by assuming the position of research director of the San Diego Zoological Garden, but he returned to his original profession afterward, again keeping up with both fields.

Kurt is one of the most energetic and enthusiastic persons I know, and his enthusiasm is highly contagious.

While he was at Dartmouth, practically every member in the Pathology Department did some work on chromosomes. He imposed this enthusiasm on me: the collaborative work with him in compiling karyotypic data of mammals into *An Atlas of Mammalian Chromosomes,* which was published by Springer-Verlag as a multivolume series.

Most biologists are not wealthy enough to join country clubs; but one country club, the Kenwood of Bethesda, Maryland, was popular among biologists, at least during the 1960s. The National Institutes of Health constantly retain many biomedical people of various specialties as consultants for study sections, councils, and other advisory groups. When these groups held meetings, lodging became a problem in Bethesda. Most motels and Kenwood would be booked full, and each lodging place would have several groups staying at the same time. I liked Kenwood more than many motels because its restaurant had excellent food and its bar was good, although its rooms were far from fabulous. After a long day of hard work, we all enjoyed having a drink in the lounge and a chat before dinner. After dinner, many of us would return to the bar, since there was nothing else to do unless we wished to retire early.

During one of the Cell Biology Study Section meetings in 1966, I joined some friends to have a drink after dinner at Kenwood. I saw a familiar figure walking in. It was Kurt, who was in Bethesda for the Pathology Study Section meeting. Naturally, I left my original group to join Kurt to talk about chromosomes. By that time, we both had collected many karyotypes of various mammals but had not published most of them because many represented isolated species of a genus or a family. We also knew that a number of cytologists had in their drawers or file cabinets karyotypes of mammalian species that they did not intend to publish. It might be a useful service if one could pool all these to publish an atlas. This would be a tremendously time-consuming job; but Kurt and I had consumed more Johnnie Walker than we should. He suggested that we publish the atlas together. The conversation gave Kurt and me nearly ten years of hard work.

The first volume of 50 species was relatively easy. We presented karyotypes of common species, including man, most domestic animals, laboratory animals, and some zoo animals. Volume 1 was published in 1967. Unfortunately, the job later became, instead of fun, a chore. In my own case, I was winding down my research activities on mammalian cytotaxonomy in the early 1970s because the field already had a sufficient number of good investigators, and I was turning my attention to other research problems. Therefore, I did not diligently search for new species to karyotype just for the atlas. It was necessary for me to beg friends to donate their karyotypes to me in order to contribute my share to the atlas. If it had not been for Kurt's insistence, I would have stopped the series after the sixth or the seventh volume. But he urged me to complete ten for a round figure.

The conventional karyotypes became out of date after the banding techniques came into being. But I think it is unnecessary to present banded karyotypes of all species. For diploid numbers and gross morphology, the *Atlas* is still a useful reference.

12 Anomalous Sex Chromosome Systems

Modern cytological investigations have firmly established the conclusion that in mammals the male is always heterogametic, or has an XX/XY sex determination system. Many earlier reports of X0 males prove to be erroneous. The sizes of the X and the Y may vary considerably, but rarely are the two morphologically indistinguishable. Ohno *et al.* (1964) concluded that the X chromosome of the eutherian mammals constitutes approximately 5% of the genome, regardless of the diploid number. In species with numerous autosomes (such as the dog), the X may be the longest element. In species with low diploid numbers, such as many bats, the X may be among the shortest elements. Exceptional cases occur when the X chromosome carries a large amount of constitutive heterochromatin, such as the Chinese hamster and the European field vole *Microtus agrestis*. As far as the functional protion (euchromatin) of the X chromosome is concerned, the genetic content of the X in relation to the entire genome is remarkably constant.

The Y chromosome is even more variable in size then the X, from a microscopically barely perceptible minute (e.g., in the ocelot) to an enormous element (e.g., in *Microtus agrestis*). The bulk of the large Y chromosome is made of constitutive heterochromatin. Since species with a minute Y function without detrimental effects for male determination, it must be assumed that the genetic messages contained in this tiny package are sufficient. Therefore, additional chromatin can be considered extraneous. In *Microtus agrestis* the functional portion of the huge Y chromosome appears to be a tiny segment near the centromere. Probably the basic Y chromosome of all mammalian species is a minute segment.

Cytologists have found a number of anomalous cases of sex-determining mechanisms in various orders and families of mammals, and these anomalies have been classically referred to as multiple sex chromosome systems, symbolized by XX/XY_1Y_2 and $X_1X_1X_2X_2/X_1X_2Y$. These are really misnomers, however, because they give an impression that these species possess more than one pair of sex chromosomes.

The anomalous types just mentioned are the results of translocations between the sex chromosomes and autosomes. The readers can find detailed information on this subject in a review by Karl Fredga (1970), who did a considerable amount of work in this area. Of course, Fredga realizes the underlying mechanisms, but he still prefers to use the XY_1Y_2 type of designation, as witnessed by his discussions:

The sex chromosome mechanism, which equally well could be designated A^XA^X/A^XY A, has arisen by a translocation between the original X chromosome and an autosome. One centromere is lost. The homologue of the autosome is unchanged and is designated Y_2. The original Y chromosome is designated Y_1 and is as a rule much smaller than Y_2.

There are two major types of sex chromosome/autosome translocations, one involving the X and the other involving the Y. These are summarized as follows:

1. X-Autosome Translocation

When an autosome is translocated onto an X chromosome, and the homologous autosome is not translocated onto the Y, the male individual possesses an odd diploid number, one more than the female. It appears that the male has two Y chromosomes. If we use the symbol $\widehat{XA}$ for the translocation, then the female individuals have $\widehat{XA}/\widehat{XA}$ and the males $\widehat{XA}$ A Y constitutions. In meiosis the free A synapses with the A portion of the X and the Y will associate with the original X, thus forming a trivalent. A number of species distributed in five different orders (Marsupialia, Insectivora, Chiroptera, Rodentia, and Artiodactyla), possess this type of sex chromosome. This phenomenon indicates that the translocations were independent events. In the fruit bat genus *Carollia*, however, all species except one have the same translocation (Figure 12.1). This suggests that the translocation was

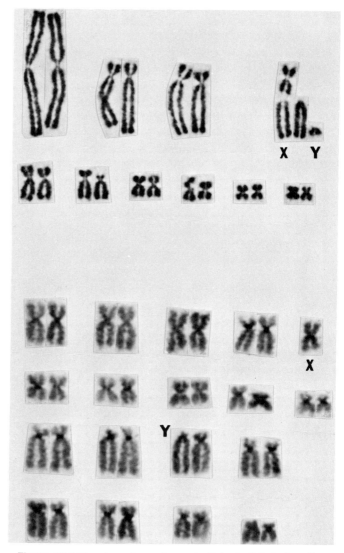

Figure 12.1 Top, karyotype of a male Seba fruit bat, *Carollia perspicillata* (2*n* = 21). The X chromosome has been translocated onto an acrocentric autosome, whose homologue remains free; bottom, karyotype of a male marsh mongoose *Atilax* (2*n* = 35). The Y chromosome (arrow) has been translocated onto an autosome. (Courtesy of Dr. Sen. Pathak.)

established before species divergence (Patton and Gardner, 1971). The lone exception, *C. castanea* from Peru, exhibits a karyotype before the translocation, with a relatively small biarmed X chromosome, a small Y, and a pair of long telocentric autosomes in both sexes (Patton and Gardner, 1971).

Perhaps the most well-known species with the $\widehat{XA}$ $\widehat{XA}/\widehat{XA}$ A Y type of sex determination system is the Tasmanian rat kangaroo (Sharman *et al.*, 1950; Shaw and Krooth, 1964).

2. Y-Autosome Translocation

When a Y chromosome is translocated onto an autosome but the X is not involved, the female will have a normal XX genotype but the male will have an X $\widehat{AY}$ genotype. The system can be symbolized as XX AA/X A $\widehat{AY}$, or the classic symbol, $X_1X_1X_2X_2/X_1X_2Y$. The male will again possess an odd diploid number, but it will have one less chromosome than the female, giving an impression of an XO constitution.

One of the earliest reported cases of Y-autosome translocation was found in the African pigmy mouse *Leggada* by Robert Matthey (1966). The basic diploid number of this genus is 36 acrocentric chromosomes, but in one species the males have only 35 chromosomes, with one metacentric chromosome representing the translocation product between the Y and an autosome.

Probably the most comprehensive study in this category was made by Karl Fredga on the mongoose genus *Herpestes*. Most species possess a diploid number of 36 for the female and 35 for the male, but other species may have different diploid numbers, with the male invariably showing one chromosome fewer than the female. However, mongooses outside of genera *Herpestes* and *Atilax* have a classic XX/XY sex determination system. When I was surveying the chromosomes of the mammals of the Houston Zoo, I once obtained a skin biopsy from a male marsh mongoose, *Atilax*. I found all the cells with 35 chromosomes and thought this mongoose had Turner's syndrome. I called the veterinarian to find out whether we made a mistake taking a biopsy from a female instead. He was insulted: "How could I have misidentified the sex?" Then I asked him to check whether the male had testicles or not. "He has already sired two litters. Nothing is wrong with that old boy." "Well," I declared, "he is a weird one. He has no Y chromosome." At that time I was not aware of Karl's work on *Herpestes*.

Unlike the X chromosome, the Y chromosome of many mammals may be very small. When the Y chromosome is very small, it may be difficult to identify it in case of translocation. In spermatogenesis, a multivalent formation is a good indication of the existence of a Y-autosome translocation.

As in the case of X-autosome translocations, the Y-autosome translocations must have occurred repeatedly in the course of mammalian evolution, since seven orders have such representatives. Fredga's review (1970) lists five orders (13 species), and two more, one in a bat and another in a marmoset, have been added since then. However, most of these are isolated cases; but the translocation must have occurred before the species divergence in *Herpestes*.

I was greatly honored when Albert Levan asked me to be the chief examiner of Karl Fredga for his disputation—an examination system for advanced academic positions. I was told that this elaborate system was preserved only in southern Sweden, which soon will also abandon it in favor of a less stringent practice. Naturally, I did not know all the traditions and Albert did not tell me in any detail. For example, I should have donned my tuxedo during this formal affair, but did not. I should have conducted the examination for at least four hours but I finished my part in three. I should have a prepared statement to be read to Karl (Pardon me, it should be Dr. Fredga) on how pleased I was with his performance and his knowledge and offered congratulations. I did offer my sincerest congratulations, but I simply said "Karl, you did a great job." In the evening, the candidate is supposed to throw a big dinner party with lots of skols. Everyone was in formal attire except the chief examiner at the head table. In a forced impromptu speech, I blamed Albert for not letting me know the custom in advance, or I would have brought my complete cowboy outfit, which is considered formal attire in Texas. Luckily, no harm was done, as Karl subsequently was appointed to the chair held by Albert who retired.

The sex chromosome/autosome translocations represent only modifications of the classic XX/XY sex determination mechanism and are, therefore, not difficult to explain. However, in a few mammalian species, the sex determining mechanisms are either complicated or are not known. The situation existing in the creeping vole (*Microtus oregoni*) is unique. Robert Matthey (1958) reported 17 chromosomes in the germ cells of the male and in the somatic cells of the female. This was reexamined carefully by Susumu Ohno and his co-workers (Ohno *et al.*, 1963, 1966), who found that both sexes of this species are gonosomic mosaics. In the male the somatic cell has 18 chromosomes (XY), but the germ cell has only 17 (0Y). A predirected nondisjunction of the sex chromosomes occurs in primordial germ cells in fetal testis, giving rise to XXY and 0Y germinal cells. Only the 0Y line differentiates into spermatogonia. Thus, the X chromosome of all individuals invariably comes from the mother, but the female zygote must start with 17 chromosomes (X0). In primordial germ cells of fetal ovaries, nondisjunction of the X chromosome takes place, giving rise to XX and 00 lines, and the latter perishes. Thus, in the female the somatic cell has 17 chromosomes (X0) and the germline has 18 chromosomes (XX). The two X chromosomes are sisters instead of homologs.

The mole vole *Ellobius lutescens* is more a mystery. Both sexes have 17 chromosomes in somatic as well as in germinal lines. The odd chromosome is morphologically identical in both sexes (Matthey, 1958). There is no sex chromatin in female cells. Presumably, this odd chromosome is the X, but it is not known whether the species actually has a Y, or if it does, where.

13 The Somatic Cell Genetics Conferences

In 1962, I was invited to present a seminar and to stay for a few days to visit various laboratories of the Biology Division of the Oak Ridge National Laboratory. I was pleased to make that trip to one of the centers of modern biological sciences. The laboratory is located a few miles outside Oak Ridge, but I stayed in a motel in town. Every morning Michael Bender and Ernest Chu came to pick me up to go to the laboratory. It was one of the most enjoyable visits I have had. One morning, on our way to the lab, I commented that we should have this type of gathering more often, perhaps with a few more people to stimulate discussion and exchange of ideas. Mike and Ernie asked me whether I knew of the *Drosophila* meetings. I said no. They told me it was unintentionally started by Dan Lindsley, then also working in Oak Ridge.

According to Dan's own recollection, he, Larry Sandler, and Bill Baker had a good time together at the International Genetics Congress in Montreal and agreed to have an extended bull session the next fall in Madison, Wisconsin. The next year Dan went to Chicago to give a seminar, and he and Bill Baker (incidentally, Bill and I also shared the same major professor) drove to Milwaukee to pick up Ted Pittenger and then drove to Madison to meet Sandler. There they sat and talked about their latest results. As Dan remembered, James Crow and probably some others also joined this informal meeting. The experience appeared most rewarding to all participants and they decided to have another one the following year in Chicago. The word got around, and several more people attended the second year.

But the reason the *Drosophila* meeting became so big later on was because of a government regulation. In order to get his expenses paid by the institute, Dan had to make a fictitious meeting called the "Midwestern *Drosophila* Research Conference" to satisfy the administrators and the auditors. The Biology Division of the Oak Ridge National Laboratory used to publish a weekly bulletin, including a list of staff travel activities. When Lindsley's attendance at this Midwestern Drosophila Research Conference was announced, many Drosophila geneticists were offended because they had no idea such a conference existed. So Dan received a number of complaints and had to hold an open meeting in Oak Ridge. About 60 people came. Recalled Lindsley,

The conference was run like a Quaker meeting, with people getting up to talk as the spirit moved them. Things moved slowly at first, but as people got warmed up we couldn't shut them up. Now the meeting is an annual affair with concurrent sessions, social events, and all the accouterments of a large national meeting.

I guess this was one of the few good things bureaucracy indirectly created.

Anyway, Mike, Ernie, and I decided that we would try a chromosome conference. Mike and Ernie talked to Dr. Alexander Hollaender, then the division chief, to see if Oak Ridge would host such an informal gathering. Hollaender graciously agreed. In the fall of that year, when the gorgeous autumn colors were at their peak, the first Conference on Mammalian Chromosomes and Somatic Cell Genetics was held at Gatlinburg, Tennessee, a lovely place in the Great Smokey Mountains. About 70 people met for three days, with one afternoon free for mountain climbing, sightseeing, or just relaxing. I remember that it was the first time I had met a number of human geneticists and cytogeneticists, including Margery Shaw, O. J. (Jack) Miller, and James German.

Everyone was pleased with the meeting and decided to hold another one the following year. The participants elected Margery to chair the next year's conference, and decided to have no official organization. This tradition has been maintained ever since: The participants select a person to organize the next meeting and this person elects a few friends to form a committee, which has the authority to run the conference. The name of this conference, however, has been shortened to the Somatic Cell Genetics Conference.

I did not attend this conference series every year, but I did go to it several times. I believe the one I enjoyed most was the 1964 meeting in San Juan, Puerto Rico, at the Condado Beach Hotel, which was closed several years later. The setting was exotic, the rum drinks were splendid, and the discussion, lively. The chancellor of the University of Puerto Rico threw a party for us one evening at his mansion, with students donning ancient Spanish costumes and singing sixteenth-century Spanish songs. I had only one regret. My old chief, Charles Pomerat, had died in the spring of that year: so a sad feeling possessed me every now and then, particularly when I was asked by the Organizing Committee to present a eulogy. It was the first time ever I had read from a prepared manuscript because I knew I could not make an impromptu speech of this kind. Since this eulogy was never published officially, I append it to this chapter as a souvenir.

CHARLES MARC POMERAT, 1905–1964

Awed by his ability and power of persuasion, I once facetiously mentioned to Dr. C. M. Pomerat that when he died I would write his obituary. It was to be entitled "Charles Marc Pomerat, the man who could sell refrigerators to the Eskimos." It didn't occur to me that the jest was a doleful prophecy, for after his untimely death, only a few years later, I was asked to write his eulogy.

I have no intention of presenting a chronological account of Pomerat's life; I leave that to biographers. Rather, I would like to describe, from what I know and from what I personally encountered, a remarkable personality; a man of multiple talents and warm heart, loved and perhaps also hated by many persons.

For four years, first as a postdoctoral fellow and later as a colleague in his laboratory, then at the University of Texas Medical Branch in Galveston, Texas, Dr. Pomerat trained, disciplined, encouraged, and assisted me in various ways. Among the persons who have changed or directed the course of my life, Pomerat was one of the most significant. When I first went to his lab, I had no background in tissue culture and related fields, for I had been trained as a *Drosophila* geneticist. In an interview prior to my taking the fellowship, he impressed me as a great orator as well as a great scientist. He persuaded me to take a chance to see what can be done with the dreadful, hopeless chromosomes of mammalian cells. If I have made any contribution to the field of mammalian ctyogenetics and cell biology, it was Pomerat's foresight that promoted it.

Tissue culture was not a popular tool in biological research before 1950. The techniques were tedious and the requirements, stringent. Most investigators using tissue culture techniques were morphologists and nutritionists. Because of the difficulties in procuring material and maintaining supplies, some tissue culturists even discouraged neophyte enthusiasts to take on such a task. Pomerat talked about Carrel who had glorified tissue culture but also buried it because he regarded tissue culture as a religion. Pomerat's attitude was to let as many persons use this material as possible, so that we could gain more knowledge. He often said, "So what if we don't have a pure cell culture. We'll use mixed cultures and study cell ecology." He tried to sell everyone on the idea that growing human or mammalian tissues *in vitro* was easy. Day in and day out, his laboratory was filled with visitors hopefully watching technicians doing various routine chores and photographers taking pictures, and, of course, listening to Pomerat's preaching. Numerous investigators from over the world asked to stay in his lab for a period of time. He never refused them. Once, the lab (not a big one) had twelve persons holding doctor's degrees, each with a drawer to house his belongings. Incidentally, the tissue culture laboratory was situated in the Texas State Psychiatric Hospital. It was rather fitting because we often fondly referred to the lab as "The Mad House."

Pomerat was a born teacher; he could not think of refusing anyone who desired to learn, no matter how crowded his place was and how crowded his schedule was. It was this magnanimous spirit that deeply moved me; I have not done so well in this respect. Pomerat also talked about his research ideas freely. When someone cautioned him that he should keep these ideas secret until he had the results "in the bag," he replied, "So let them steal my ideas. I'll get some more."

Trained as a histologist in Harvard, Pomerat learned tissue culture from Professor E. N. Willmer of Cambridge University. He taught at the University of Alabama for three years before settling down in Galveston in 1943. He tried to push research in all directions with cells in culture. Perhaps people criticized him as superficial. Such criticism was only partially true. His main objective was to urge everyone explore his own field as deeply as possible so that Pomerat himself could find time to educate someone else. In other words, Pomerat was a super salesman, selling all his knowledge and inspiration to others for them to go into the various gates of the research frontier. Nevertheless, he himself did have a research program which he cherished most, viz., experimental neurology.

In his heart, Charles Pomerat was more an artist than a scientist. He painted diligently—water colors, oils, etchings, sketches. One time a New Orleans man offered him $250 in exchange of four water colors which he could execute in two afternoons. He declined, saying that art would be distasteful when it became a business. Rather, he would patiently explain to a 7-year-old girl the basic principles of art. One sunny afternoon he took this little girl to a weed-sprawled, deserted alley and led her to an appreciation of beauty in squalor, by demonstrating a complete landscape painting. This insignificant episode profoundly inspired the little girl, who at the age of twelve won a Houston Art Museum scholarship and proceeded to learn art without interruption. This girl is my daughter.

Pomerat's artistic achievement was not limited to painting. He was an excellent amateur photographer and a good cook, particularly of exotic foods. He was an authority on Gothic architecture. In fact, even his scientific endeavors carried a heavy artistic touch. The beauty of his motion pictures of cell activities is so well known that no further mention is necessary. He would spend hours with photographers searching for a perfect sequence. Yes, Pomerat was a perfectionist. He attended every detail in every manuscript, exhibit, or lecture. People became hypnotized by his oral presentations, thinking he was a gifted speaker. He was; but seldom did they know that he had put hours of preparation and recitation into every sentence he so eloquently delivered to his audience.

It would be unfair to Pomerat as well as to his memory if I do not mention his shortcomings. He was a temperamental person. When he was in a bad mood, especially when under pressure, anyone who happened to be caught in his eyes might be the target of an explosion. Generally, the storm abated quickly, and remorse invariably overpowered him. Would his kindness be enough to compensate for his faults? To most persons, the answer would be yes; for people forgave him knowing that he did not mean what he did.

Early in 1960, he underwent a colostomy, because of an adenocarcinoma, at the John Sealy Hospital in Galveston. He recovered quickly, and regained his lively spirit. In the same year, he left the University of Texas to take a position in the Pasadena Foundation for Medical Research, Pasadena, California. Probably realizing that his life was limited, he worked harder and played harder than ever. He trotted over the world, lectured constantly, and painted ferociously. In 1962, when the first Mammalian Chromosome Conference was held in Gatlinburg, Tennessee, Pomerat was a spark plug of the meeting.

Everyone who knew Pomerat mourns his death, including the little girl, who is no longer a little girl. A few weeks before he died, she mentioned to me, "I wish Dr. Pomerat would visit us and see my paintings." She will not be able to see him anymore, nor will anyone else, but Pomerat's influence will be felt for decades to come.

14 Mammals for Cytogeneticists

In years past, laboratory animals were bred for various purposes, but none for their good cytological characteristics. The Chinese hamster, *Cricetulus griseus* (Figure 14.1), was no exception. It became a laboratory animal because of several morphological features, parasitic relations, and disease susceptibilities. It took many years of observation and experimentation, both in China and in Europe, for this species to be established in laboratory colonies. Matthey (1951) determined its diploid number as 22, one of the lowest known among eutherian mammals.

According to George Yerganian (1957), who gave a brief historical account of this species as a laboratory animal, a man named Schwentker first successfully bred the Chinese hamsters on a commercial basis. Yerganian's own colony and probably many others in the Western world came from this stock. I believe George was the first person in the United States to start breeding the Chinese hamsters. He did so while taking a postdoctoral traineeship in the early 1950s under the supervision of a plant cytogeneticist. He became so engrossed with his little creatures that he spent less and less time doing research on plant chromosomes. He saw the possibilities of using the Chinese hamsters for many research problems, particularly in genetics and cytogenetics, and patiently observed the mating behavior in order to work out an effective breeding system for creating inbred lines. These developments, however, upset his supervisor, who finally gave George an ultimatum: "Either you go or the

Figure 14.1. A Chinese hamster.

Chinese hamsters go.'' Yerganian's reply: ''We both go.''

In a way, I do not assail the supervisor for his action. After all, plants were THE thing for chromosome studies at that time, and very few persons had the foresight and vision to predict that the little creatures from northern China had a great future. On the other hand, I admire George not only for his conviction, but also for his defiance. If it were not for him, we might have missed one of the world's most useful materials for chromosome studies.

So Yerganian and the hamsters finally relocated at the Children's Cancer Research Foundation in Boston. He became the chief breeder and advocate of the Chinese hamster in the United States in the 1950s and early 1960s. His wife, Sona, frequently traveled with him to scientific meetings. Sona is a pianist. According to Sona, when George proposed marriage to her, he thought that as a scientist his life could be enriched by her music, and as a musician, her life could be embellished by a touch of science. But she concluded: ''In the last ten years, we ate, sat, and slept with the Chinese hamsters and chromosomes!'' George, sitting by her side, simply grinned.

In 1960, when Carolyn Somers and I were working on the effects of bromodeoxyuridine on chromosomes, we decided that we must obtain a cell line of the Chinese hamster in order to get more precise data. So I wrote to

George for help. He kindly shipped me two of his near-diploid cell lines established from a female animal. Carolyn and I immediately utilized the cells for our study and happily discovered the nonrandom distribution of chromosome damage induced by this thymidine analog. We interpreted it as a suggestion of nonrandom distribution of DNA base pairs along the chromosomes (Hsu and Somers, 1961). This interpretation received mixed reactions until some ten years later, when more definite evidence was found.

The Chinese hamster cells soon became the princesses of my laboratory; we designed many experiments utilizing these cells. However, we soon needed some live animals for other experiments. George sent me some animals, but they failed to breed. I then heard that Michael Bender at Oak Ridge had a good colony going, using stock animals from England. Mike came to Houston for a visit and personally carried three pairs of young breeders, from which we maintained our little colony for many years.

One of the males, however, was required for the initiation of cell lines. One weekend I took my daughter, Margaret, to my laboratory. One of the technicians who took care of my animals and worked on Saturdays tried to entertain Margaret by showing her our new animals. Not unnaturally, Margaret reached into a cage to play with one. He bit her finger. Margaret told me that this hamster was as mean as one of her classmates, Donald. Thus this animal received the name, Donald. Since I was hoping to get a colony of docile animals, Donald was the logical choice to be sacrificed as the tissue donor.

His lung cultures grew luxuriantly, with excellent diploid chromosomes. Through the years, we have distributed this cell line to scientists over the world, including the American Type Culture Association. The symbol for this line, Don, is the shortened version of the animal's name. It is not DON as it has appeared in some scientific papers. Before the discovery of the unique Indian muntjac, the Chinese hamster chromosomes were the best eutherian mammals could offer—low diploid number, great size difference, and unequivocal recognition of several pairs by metaphase morphology alone.

Since we were using the Chinese hamster cells for a variety of research problems, we decided to construct an idiogram for this species (Hsu and Zenzes, 1964). Being a *Drosophila* geneticist, I was accustomed to the nomen-

clature system that does not single out the sex chromosomes. In *Drosophila melanogaster,* for example, the sex
chromosomes are placed in the number 1 position. In the
human chromosome system (Chapter 7), the sex chromosomes are not numbered. Maria Zenzes and I placed the
sex chromosomes of the Chinese hamster as the No. 3
pair.

In the meantime, George was interested in another
species of small hamster, *Cricetulus migratorius,* a native
of Armenia and Turkey. Since both George and Sona are
of Armenian ancestry and speak the language fluently,
they made arrangements to visit the U.S.S.R., particularly their homeland Armenia. With the help of Russian
zoologists, George trapped some specimens of this particular hamster species. However, the Yerganians went
through a good deal of hardship trying to carry the live
specimens out of Russia. Finally, George threatened to
leave the hamsters in Russia and go across the border to
catch some in Turkey; he would then call the animals
Turkish hamsters instead of Armenian hamsters. Somehow this worked, and George has a colony of this species
in his laboratory. Unfortunately, the chromosomes of the
Armenian hamsters are no better than those of the
Chinese hamsters. In a meeting, George viciously
criticized my nomenclature system of placing the sex
chromosomes in the numerical lineup. To me it was a
trivial point, but at the urging of friends in the audience, I
finally retorted that George should mind the Armenian
hamster business and leave the Chinese hamsters to a
Chinese.

I must emphasize that despite many virtues, the
Chinese hamster chromosomes are not perfect. The X
chromosome is not always separable from the autosomal
pair No. 4, several pairs of autosomes are morphologically similar, and the two arms of the near metacentrics
(Nos. 2,5,10,11) are also indistinguishable. As biomedical
investigations expanded to various directions, cell
biologists and cytogeneticists quickly found that no one
material is ideal for answering all questions. For example, some experiments may require fast-growing cell
populations but others, slow-growing cell populations
with a long G_1 phase. Cytogeneticists may need to employ animal cells with numerous acrocentrics, with a very
small Y chromosome, with a large amount of heterochromatin or practically no heterochromatin, and many
other features. Therefore, the Chinese hamster cells

alone cannot satisfy all these requirements and cytologists continued to look for materials with different properties. My survey of mammalian karyotypes (Chapter 9) was, at least in part, an attempt to reach this goal.

Most of the Australian and Tasmanian marsupials have a low diploid number. Several groups of cytologists tried to procure the Tasmanian rat kangaroo (*Potorous tridactylus*), which has a diploid number of ♀12 and ♂13 (McIntosh and Sharman, 1953). A detailed karyotype of this species was described by Shaw and Krooth (1964). The Naval Biological Research Laboratory at Oakland finally established two kidney cell lines, which were subsequently preserved in the American Type Culture Collection. The diploid number is indeed very low, but the chromosome morphology is not better than the Chinese hamster and it is extremely difficult to obtain live animals unless one works in Tasmania.

Another seemingly promising material was the creeping vole *Microtus oregoni* ($2n = ♂$ 18, ♀ 17), which has fantastic chromosomes and unique sex determination mechanisms (Chapter 12). Susumu Ohno sent me several specimens of these animals as a gift in the early 1960s, but only one survived in transit. These creatures are not only difficult to procure but are also difficult to keep in captivity.

Then there is the European field vole *Microtus agrestis* ($2n = 50$), which has huge sex chromosomes with a large amount of heterochromatin. But its autosomes are small and morphologically poor. In Europe, the animals can be trapped in the field; but in the Americas it is necessary to maintain a colony or to keep vigorous cell lines. Prior to the invention of banding techniques, a number of cytologists coveted *Microtus agrestis* for its unusual sex chromosomes. I was one of these enthusiasts, and I tried hard to get a pair or two to start a colony. Finally, Jorge Yunis gave me a pair. To express our appreciation, one of my technicians named the male George and the female, Eunice. Later, Bruce Cattanach gave me a male specimen (naturally, we named him Brucie) who sired many offspring. At one time, I had over 300 animals, but they were wiped out within one week when my institute hired an exterminator to spray some insecticide in the animal quarters. The insecticide did not harm the Chinese hamsters or *Peromyscus*, but the voles succumbed. I was so heartbroken that I never revived my interest in *Microtus*. Actually, after the invention of the C-band technique for the demonstration of constitutive heterochromatin (Chapter 18), a number of good materials have been discovered and interest in *M. agrestis* has waned somewhat.

Although cells in culture cannot completely replace intact organisms in some areas of biomedical research, they offer excellent materials for others. In fact, in many research programs, utilization of cell cultures is the only logical and economical approach. For example, the Indian muntjac cell lines established by Doris Wurster have been extensively used by cytogeneticists and somatic cell geneticists. Think what it would entail if one had to maintain a herd of the muntjacs in order to perform experiments. Improvements in cell culture and cell preservation methods have made great contributions to a variety of branches of biological science, including cytogenetics. Thus, when a particular animal species is considered useful in one respect or another by a biologist, the first thing to do is to establish a cell line as experimental material and freeze the live cells for future use. More and more molecular biologists are also beginning to use cell cultures. The kangaroo rat *Dipodomys ordii* has a large amount of satellite DNA. David Prescott established cell lines from these rodents for his own study, but also gave seed stocks to other interested investigators. In my laboratory, we established a number of cell lines from the deer mice (genus *Peromyscus*) for its repetitive DNA and heterochromatin and utilized them for a variety of research problems. Likewise, we distributed them to other investigators for their studies.

As biomedical research becomes more and more diversified, the use of special animals by geneticists, immunologists, virologists, physiologists, biochemists, and molecular biologists will steadily increase, and cell lines of many animals will likewise become more popular. Cytogeneticists were just a bit ahead of others.

15 Old Cultures Never Die?

Although Wilton Earle established the permanent cell line strain L from a C3H mouse as early as 1940, few biologists took advantage of such material for research until the next decade, when, in 1951, George Gey established the HeLa line from a human carcinoma. One of the reasons biologists were not more interested in cell and tissue cultures was that it was both difficult and expensive to set up a tissue culture laboratory. The interest in cell culture at the beginning of the 1950s was stimulated by the pioneering work of John Enders and his associates, who claimed that cells in culture could accept animal viruses as a substrate, thus avoiding inoculations from animal to animal and lessening the danger of accidental infection.

Virologists of this early stage were mainly concerned about growing the cells in culture and infecting the cells with the viruses in question. They seldom considered the host cells: structure, physiology, genetics, and responses to viruses. This probably was a logical development: First thing first. Virologists contributed to the field of cell culture, positively or negatively, by initiating a number of long-term cell lines from a variety of tissues and a variety of animal species. Many of these turned out to be contaminated L cells or HeLa cells (Chapter 5).

In the meantime, cytogeneticists began to analyze the chromosomes of the available long-term cell lines. Both HeLa and L were found to be heteroploid. It was no surprise to find the HeLa cells with abnormal karyotypes, because it was already well known prior to that time that neoplastic cells had abnormal chromosome constitutions.

But the L cells, and some other mouse cell lines established by Katherine Sanford in Earle's laboratory, also exhibited heteroploidy with stemline numbers in the fifties, sixties, or seventies. Many chromosomes were biarmed, indicating centric fusions, since all the normal mouse chromosomes are telocentric.

These observations raised a question: Do cells of normal tissues invariably change their chromosome constitution after grown in culture for a period of time? The answer was affirmative insofar as many rodent cell lines were concerned. Diploid Chinese hamster cells were abundant in the primary and early subcultures, but slowly, or sometimes suddenly, aneuploid cells increased in the cell population and eventually replaced the diploids. A similar situation was found in the mouse and some other species. However, many tissue culturists experienced difficulty in maintaining long-term cultures of human cells derived from normal tissues. The cells would grow vigorously for a number of subculture generations (passages) and then slow down and finally cease to proliferate and die.

This phenomenon was extensively studied by Leonard Hayflick and Paul Moorhead (1961). They initiated cell cultures from a variety of normal human tissues and kept records on growth activity, subculture intervals, passage number, and chromosome picture. They found that the history of a human cell line could be divided roughly into four stages, or phases. Phase I is the initial stage when the cells adapt to the *in vitro* environment. During this stage, the growth rate is relatively slow, but the cells gradually pick up their growth potential for the Phase II of exponential growth. This phase lasts until the late twentieth or the early thirtieth passages and the growth rate slows down. In Phase III, the passage time becomes progressively longer. During fortieth and fiftieth passages (Phase IV), the line completely ceases proliferation; the cells become large and granular; and eventually they degenerate.

Hayflick and Moorhead also found that the cells maintained a predominantly diploid constitution even toward the end of the culture's life. Thus, they concluded that human diploid cells have a finite life span *in vitro* and that, unlike the cells of some rodents, the chromosomes of the human cells do not alter their constitution. In general, 40–50 passages is the limit of the *in vitro* life. Hayflick followed with several papers to propose his hypothesis of cell senescence: Human cells have a built-in, or programmed, capacity for multiplication, and when that capacity reaches its end, senescence is the result. Cell culture really exploits to the fullest this programmed capacity for regeneration.

Attempts by Hayflick and Moorhead to freeze-store cells at various passages and revive them later showed that storage did not alter the trend toward senescence. All the cultures, regardless of the passage number at the time of storage, went to a cumulative total of 40–50 passages and succumbed.

During the early 1960s, a practical consideration arose. Production of human viral vaccines became an important endeavor in public health. One of the problems to be resolved was what cells should be used as substrate for viral propagation. Several conferences were held, and standardization guidelines were proposed.

The consensus was that diploid human cell lines should be used. Despite the senescence phenomenon, one cell line can produce an astronomical number of cells for industrial use if cells in early passages are frozen for future revival. Len Hayflick's human cell line WI-38 was selected as the candidate line and, in the meantime, other candidate lines were to be developed, characterized, and approved before the WI-38 cells were eventually exhausted.

One of the developments in long-term human cell cultures in the late 1960s was the discovery of lymphoid cell lines. Cytogeneticists had been using human blood or lymphocyte cultures as a routine procedure for chromosome analyses. These cultures usually multiply rapidly for a few cell generations and then stop. No one was able to obtain a long-term cell line from lymphocytes of normal persons, except in a couple of cases reported by George Moore and associates (1967). One of these lines, when examined by Christofinis (1969), was found to be predominantly diploid up to 130 passages. However, prior to that time, some indications existed that bone marrow cells from patients suffering from infectious mononucleosis may transform in culture into a lymphoblastoid form and become permanent cell lines. A systematic study was later carried out by Glade et al. (1968) to test whether peripheral lymphocytes of patients with infectious mononucleosis can be cultivated in vitro for long periods of time. The data presented by these investigators showed that most of these cultures could be maintained *in vitro* as permanent suspension cultures. Again, diploidy prevailed.

Thus, the Epstein–Barr virus may have something to do with the ability of the lymphoid element to proliferate in culture. Arthur Bloom and Kurt Hirschhorn contributed a good deal of their efforts in this direction, and found that if the cells of a long-term lymphoid cell line were lysed and an aliquot of the medium filtrate was added to a lymphocyte culture from a normal person, the chances of establishing permanent lymphoid cell lines increased. Later, Kurt found that a certain brand of phytohemagglutinin could do the same without resorting to using the cell lysate. In all cases, diploidy was maintained. Actually, as I was told, the lymphoid lines are invariably the descendents of B-cells. Phytohemagglutinin stimulates only T-cells. When proper procedures are followed, lymphoid cell lines can be established without the use of any mitogen.

No comprehensive cytogenetic data are available on the lymphoid cell lines, but the general impression is that all of these have some minor rearrangements, even though the diploid number is maintained. Whether the minor changes in the chromosome constitution causes the change in the physiology of the Epstein–Barr-infected cells to "transform" remains a question, but it appears that diploidy per se is not the a priori reason for cellular senescence. In fact, WI-38 or other so-called diploid fibroblast lines are not devoid of aneuploid elements or cells with rearrangements.

The human lymphoid lines have many advantages over fibroblast cultures. First, the lymphoid cultures can be continuously propagated with no sign of reaching a senescence stage. Second, they are suspension cultures and are therefore easy to handle. Subculturing requires no scraping or trypsinization. Third, they usually grow fast. It is relatively easy to obtain large quantities of cells. Finally, for establishing long-term cell cultures from persons with genetic defects or variations, peripheral blood samples are certainly less objectionable than skin biopsies. It is true that the lymphoid cells may harbor certain viruses, which may be undesirable for certain experiments and may be more hazardous, but the virtues far outweigh the shortcomings.

Many biologists seem to accept the cell senescence hypothesis. Personally, I feel that other possibilities may also explain the phenomenon of slow deterioration of diploid human cell lines. In discussing diploid cell lines, I once commented (Hsu and Cooper, 1974):

> Everyone engaged in cell culture has encountered occasions when cell lines grow well for some time but suddenly become "sick" almost simultaneously. Usually this phenomenon is associated with a new batch of medium, new trypsin solution, or other occurrences. Glassware washing, the purity of distilled water, routine handling system, and many other factors all contribute to the successful maintenance of diploid cell lines. Earley and Stanley found that the kinds of growth medium severely affect the capability of producing aneuploid lines of rabbit cells. They found that rabbit cells can maintain diploid composition in medium without lactalbumin hydrolysate for 200 passages. We found that diploid Chinese hamster cell lines would exhibit numerous cells with aneuploidy and/or chromosome aberrations after they recover from a short period of sluggish growth. Undoubtedly the period of sluggish growth was the result of certain defects in the culture system. Perhaps other laboratories have had similar experience. Perhaps no contemporary cell culture laboratory can claim that its cell culture system represents the ultimate perfection. Then it is exceedingly probable that the cells repeatedly receive mild traumas during extended propagation *in vitro*. When the seemingly healthy cells slowly decrease their growth rate, possibly past injuries express their cumulative impact. We therefore believe that senescence of cell cultures is at best an alternate hypothesis. It is possible to think that diploid cell lines can grow forever when the ideal conditions for their cultivation are met.

Old cultures never die. They just get poisoned or contaminated.

16 The Bandwagon

In 1962, soon after some startling discoveries had been made in human cytogenetics, I wrote a review chapter on genetic cytology of mammals for a book that was published three years later (Hsu, 1965). I commented in that paper that the honeymoon was over. Here is what I wrote:

> The law of diminishing returns also applies to scientific research, because sensational cases become fewer and fewer as more and more discoveries are made. Sooner or later a number of investigators will leave this field, either discouraged because nothing exciting comes their way, or bored because the problems are not as interesting as they thought, or forced because the grant is cut. . . .

Cytogeneticists must be able to make detailed analyses of chromosomes and segments of chromosomes. This was not feasible then because many chromosomes in the human karyotype are indistinguishable by conventional staining. The only chromosomes that could be identified unequivocally were Nos. 1, 2, 3, 16 and the Y. Attempts to use DNA synthetic patterns, computer automation, and other techniques failed to offer substantial improvements in differentiation of chromosomes. Comparative cytogenetics of mammals encountered the same problem, and the entire field indeed appeared stagnant near the end of the 1960s. If it were not for several very significant advances made in the very late 1960s and early 1970s, many cytogeneticists probably would have abandoned the field altogether.

But a second honeymoon started, almost exactly ten years after the discovery of 21-trisomy in man. In fact, several important techniques clustered within approximately a two-year period. These techniques not only revolutionized cytogenetics but also shed much light on the problem of chromosome organization. It is interesting to note that while mammalian and human cytogenetics was in its infancy, plant cytogenetics and insect cytogenetics had reached their heights of development. The situation is now reversed. What a difference a few years can make!

In the late 1960s, the late Sidney Farber, then director of the Children's Cancer Research Foundation in Boston, asked Professor T. Caspersson (Figure 16.1) to serve as a consultant to the Foundation. Caspersson, of course, has been an authority in fluorometry and interferometry for many years and his laboratory in the Karolinska Institute at Stockholm has a worldwide renown. As a consultant, Caspersson felt he should make some contribution to the institute that retained him.

Many fluorochromes can stain chromosomes and show fluorescence under ultraviolet optics. Usually the fluorescence is uniform throughout the lengths of the chromosomes; but Caspersson thought if one could attach a chemical compound such as an alkylating agent to

Figure 16.1. T. Caspersson. (Courtesy of Dr. U. Arnason.)

a fluorochrome molecule, the alkylating agent might crosslink guanines of the DNA in the chromosomes and the stain would give fluorescence. Then if the distribution of base pairs along a chromosome is nonrandom, the guanine–cytosine-rich (GC–rich) areas should receive more crosslinking of this compound molecule than adenine–thymine-rich (AT–rich) areas, consequently showing brighter fluorescence. Therefore, if all these ifs were true, one would expect a chromosome to show differential fluorescence along its length into bright and dim zones. And if this proved to be the case, one should be able to differentiate chromosomes of similar gross morphology by their fluorescence patterns.

Caspersson persuaded Ed Modest, an organic chemist at the Children's Cancer Foundation, to synthesize such a compound. Ed reluctantly tried and finally succeeded in synthesizing quinacrine mustard (QM), which was shipped to Sweden. The first series of observations with QM was done on plant chromosomes (*Scilla, Vicia,* etc.) by Lore Zech, Caspersson, and associates. They found what they had hoped to find: differential fluorescence along each chromosome. The metaphase chromosomes showed characteristic fluorescent segments (or bands) of various degrees of brightness (Caspersson *et al.*, 1969). Thus, chromosome pairs with identical gross morphology could be recognized individually by their fluorescence banding patterns.

This discovery prompted the Caspersson group to try the next logical step, examining the human chromosomes. The human chromosomes also gave them what they had hoped for: Every pair could not be recognized by its characteristic fluorescence pattern (Caspersson *et al.*, 1970). The combination of the position, the width, and the brightness of the QM fluorescence bands is unique for each chromosome so that every element in the human complement can be recognized with relative ease. This discovery, the beginning of the bandwagon, was quickly confirmed and detailed by a number of investigators, and the advance caused the existing field of human cytogenetics to become obsolete almost overnight, because numerous unsolved problems could now be solved. Figure 16.2 shows a fluorescent banding of human chromosomes.

In 1971 a number of human cytogeneticists held the Fourth Human Chromosome Standardization Conference in Paris. In this particular meeting, now known as

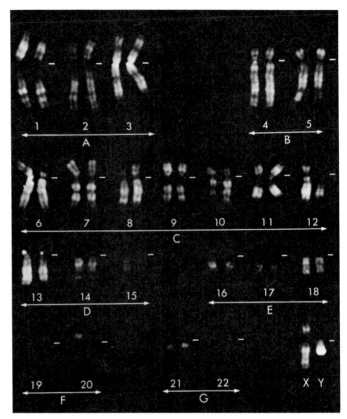

Figure 16.2. A Q-banded human karyotype. (Courtesy of Dr. C. C. Lin.)

the Paris Conference, many new discoveries were discussed, and many recommendations were made. The fluorescence banding is now known as the Q-band, recommended by the Paris Conference (1971).

Besides contributions to human cytogenetics, QM fluorescence gave an immense dividend to the chromosome studies of the laboratory mouse, *Mus musculus*. The diploid number of the mouse is 40, all telocentric. The sizes of the mouse chromosomes form a smooth gradation, and it was simply not possible to recognize any pair except the smallest (Pair 19) and the Y.

Mouse geneticists had produced numerous translocation and inversion stocks and identified their linkage group involvements by genetic methods. These were not cytologically verified, however. Only the translocation produced by Charles Ford, the T_6 marker, in which the rearrangement was so unequal that one of the component chromosomes was a minute, was advantageously used for transplantation and other experiments. Now, with QM fluorescence, the impasse suddenly opened into a wide boulevard. Here again, every mouse chromosome had its characteristic fluorescence banding pattern, and an ad hoc committee promptly recommended the adoption of the nomenclature system of the mouse chromosomes according to the Q-band patterns (Committee, 1972).

Previously, the linkage groups of the mouse were assigned according to the chronology of discovery. Since the mouse chromosomes were individually unrecognizable, there was no way to correlate the linkage groups with the chromosomes. Now the living stocks carrying reciprocal translocations could be effectively used for chromosome assignment. Suppose two different sets of translocations involve a common linkage group. Cytologically, one chromosome should show, in the heterozygotes, abnormality in both translocations. This chromosome must be equivalent to that common linkage group. Using this method, many linkage groups were cytologically identified and the break points of many translocations have been located (Miller and Miller, 1972).

The phenomenon of differential fluorescence along metaphase chromosomes not only gave a powerful tool for positively identifying individual elements and subdivisions of these chromosomes, but also fueled the study of chromosome organization. It strongly suggests that the distribution of DNA base pairs along the chromosome is nonrandom. According to Caspersson's original idea, the bright zones should represent GC-rich segments and the dim zones, AT-rich segments. From all available information, however, it is the *AT-rich* DNA that shows bright fluorescence. The mustard moiety of QM apparently does not contribute to the fluorescence banding, since quinacrine dihydrochloride (Atebrin), which has no alkylating agent attached to it, gives exactly the same fluorescence pattern. Moreover, Bernard Weisblum (Weisblum and de Haseth, 1972; Weisblum and Haenssler, 1974) found that in purified DNA solutions, the degree of brightness was correlated with the base compositions: Those with higher AT ratios gave brighter fluorescence than those with low AT ratios. Apparently quinacrine has no preferential chemical affinity for AT base pairs, but its fluorescence depends largely on the base composition.

After the invention of the Giemsa banding techniques

(Chapter 19), fluorescence banding took something of a back seat, since laboratories that cannot afford a fluorescence microscope can obtain detailed chromosome banding patterns by using a bright-field microscope. However, Giemsa banding cannot completely replace fluorescence. For example, the human Y chromosome has an extremely Q-bright distal segment in the long arm. Peter Pearson and his collaborators showed that this segment can be detected by fluorescence even in interphase nuclei, but it cannot be detected by the Giemsa banding technique (Pearson *et al.*, 1970). One can also use the intensity of fluorescence as special marker to identify variations. In addition, the Q-banding is very useful for sequential or combination photography. The chromosomes can be photographed with a fluorochrome for initial photographic records and the preparations can then be treated for various experiments, such as autoradiography or silver staining for the nucleolus organizers. The Q-bands can provide positive identification of the treated chromosomes.

Induction of good chromosome banding is not limited to quinacrine and quinacrine mustard. Alfred Gropp and his associates found some useful properties in a dibenzimidazole derivative, known as 33258 Hoechst (Hilwig and Gropp, 1972). In the chromosomes of the mouse, the centromeric areas do not show bright fluorescence when quinacrine is applied, but they are extremely bright in 33258 Hoechst preparations. This property can be advantageously utilized to differentiate mouse and human chromosomes in somatic cell hybrids. When properly used, 33258 Hoechst, which also stains AT-rich regions, does not fade as quickly as quinacrine. Samuel Latt (1973) found an additional use of 33258; namely, it can differentiate sister chromatids if a certain experimental protocol is followed (see Chapter 19).

More recently, fluorochromes with preferential affinity for GC-rich DNA have been employed for chromosome banding. These include chromomycin A_3 and mithramycin (Schweizer, 1977) and olivomycin (Sande *et al.*, 1977). Their fluorescent patterns are the reverse of those of quinacrine and 33258, that is, dull Q-bands become bright and vice versa. Drug pretreatments to chromosome preparations followed by fluorochromes are also in use to enhance resolution.

Another technical approach using fluorescence has value in molecular cytogenetics; but unfortunately it cannot be routinely adopted by other laboratories because it requires expertise in a special branch of immunochemistry. This is to use fluorescein-tagged antibodies against specific nucleosides or short base-sequences as a probe to locate chromosome areas rich in that particular base or base sequence. Most credible works came out of collaborations between Orlando J. Miller and Bernard Erlanger. As Jack Miller told me:

And now a few words about the development of immunological probes for use in cytogenetics. It arose as a result of my serving as organizer of a course in genetics for medical students at Columbia. Sam Beiser, a member of the Department of Microbiology and Elvin Kabat's first graduate student (Immunochemistry), gave several lectures in this course, and this brought us into contact. He and Bernard F. Erlanger had developed a method for producing antibodies to specific nucleosides and discovered that they would react with the haptenic base in denatured DNA. They published their findings in PNAS in 1964, and continued enlarging their battery of antisera with different base specificities. In the fall of 1965, Sam and I first discussed the possibility of using these antisera as probes of chromosome structure. Neither of us knew about the earlier studies carried out by Bob Krooth, Jo Hin Tjio, Fred Rapp, and a few others, showing that lupus antisera can bind to chromosomes, and I didn't know that David Stollar had shown that lupus sera may react much more strongly with one short nucleotide sequence than with others. If we had known about this earlier work, we might, by mistake, have been less enthusiastic about the potentials of this approach. As it was, we were both excited about the prospects, and decided to begin work on it just as soon as M.V.R. Freeman arrived. Van was an obstetrician who wanted further training in cytogenetics and research methods, and was interested in immunology. He arrived in 1967 and spent his first year immunizing rabbits and characterizing antisera. He then used a combined autoradiographic immunofluorescence method to confirm Bill Klein's demonstration that detectable amounts of single-stranded DNA are present only during the S phase of the cell cycle. He also showed that the distribution of antinucleoside antibody binding varies throughout the S period—a finding whose potential usefulness as a marker of specific points in the cell cycle has still not been exploited.

Van spent this last few months studying the binding of antinucleoside antibodies to chromosomal DNA that had been denatured by alkali. He did observe binding, but the chromosomes tended to be swollen by the NaOH treatment and he did not observe chromosome banding. Publication of his results was delayed for two years by his move to another institution, and during this time, we did no more with the antibodies. After all, nobody else was working on the problem! After Caspersson's group acquainted the world with chromosome banding, I went back over Van's pictures and noted that antibody binding was not totally even, though with swollen chromosomes the suggestions of banding were minimal. With renewed enthusiasm, my group pursued this lead. By using a milder denaturing treatment we were able to show that anti-adenosine antibodies bind specifically to the quinacrine-bright bands. Studies with antisera of other base specificities were not, at first, rewarding—in fact, they were very frustrating, because we were unable to obtain the expected reverse banding pattern with anti-C or anti-G. At this point, Bernie Erlanger played a decisive role, drawing upon his expertise as a biochemist to suggest methods of generating single-stranded regions in particular classes of DNA, e.g., photooxidative destruction of guanine residues in DNA to free up cytosine residues in the complementary strand, opening up stretches in GC-rich DNA; or UV-irradiation to produce thymine dimers in some AT-rich DNAs. One of my graduate students, Rhona Schreck, worked out the application of these methods to chromosomal DNA, with the result that we are now able to reproduce not only the general Q- and R-banding patterns but also to specifically tag heterochromatic regions of particular human chromosomes, thus greatly enhancing the potential for characterizing structurally abnormal chromosomes.

I thought over the recent history of mammalian and human cytogenetics and realized I was in the right spot at the right time on several occasions, e.g., England in 1959, studying Klinefelter's syndrome (resulting in Ford et al., 1959; Harnden et al., 1960), and New York in 1971, studying mouse chromosomes (resulting in placing the linkage groups on their chromosomes).

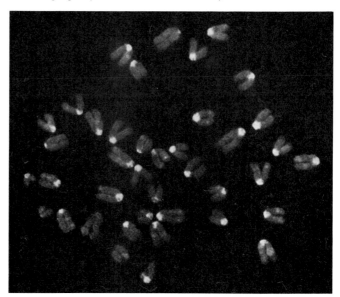

Figure 16.3. A mouse metaphase stained with fluorescent antibody against 5-methyl cytosine. (Courtesy Dr. O. J. Miller.)

Figure 16.3 shows a metaphase picture of the mouse (*Mus musculus*) labeled with antibody against 5-methylcytosine. It is known that the satellite DNA of the mouse is rich in 5-methylcytosine, and the satellite DNA is located in the heterochromatin areas near the centromeres. The cytological result confirms the biochemical data and indicates that as the immunochemical techniques improve, more information will be obtained in deciphering base distribution along chromosomes. For summary of the background the accomplishments in this area, a review paper by Miller and Erlanger (1975) is very informative.

Without question, the bandwagon will draw more crowds and its music will be sweeter.

17 *In Situ* Hybridization: Marriage Between Molecular Biology and Cytology

After the discovery that RNA molecules with sequences complementary to those of DNA can anneal with the template to form DNA–RNA hybrid molecules, biologists began to consider the possibility of DNA–RNA hybridization in cytological preparations. If a particular species of RNA is made radioactive and if such RNA molecules form hybrids with complementary DNA in the cell nuclei and chromosomes, the cytological locations of this DNA (or genes) can be identified in the autoradiographs. In principle, this procedure should be quite feasible; but in practice a number of technical problems must be solved. A number of cytologists, including myself, made such attempts but failed.

Not being a molecular biologist, the idea came to my mind relatively late, sometime in 1965. My purpose in trying *in situ* DNA–RNA hybridization experiments came purely from my interest as a cytologist in the nucleolus organizer regions (NORs). Cytologists had known for a long time that the achromatic secondary constrictions in the chromosomes are intimately associated with nucleoli (hence the term). In all probability, then, the NOR should be where the ribosomal cistrons (rDNA) are located. In the mammalian karyotypes, we noted that some species have a pair of conspicuous secondary constrictions (e.g., the rat kangaroo), some have short, multiple secondary constrictions (e.g., human), and others have no visible secondary constrictions (e.g., cattle). It would be useful to find out, by nucleic acid hybridization in cytological preparations, whether the secondary constriction regions are indeed the chromosomal locations of the ribosomal genes, and when visible secondary constrictions are absent in the karyotypes where the genes are located.

I discussed this idea with some of my colleagues, and we tried some preliminary experiments. After encountering numerous technical problems, we abandoned the project for the time being, but I kept this as one of my dreams. My dream finally came true some ten years later with help from Mary Lou Pardue (Figure 17.1), one of my favorite collaborators.

It is easy to perform *in situ* nucleic acid hybridization now, but success of the first experiment required (*a*) a lot of knowledge, both of molecular biology and of cytology; (*b*) a superior material; and above all, (*c*) a great deal of perseverence. Credit must go to Joseph G. Gall (Figure 17.2) and Mary Lou Pardue for perfecting the *in situ* hybridization procedure that brought about a marriage between molecular biology and cytology. Joe had been an authority in amphibian cytology for years, and took advantage of his knowledge and experience in this endeavor.

Let me first briefly describe the principles of *in situ* nuclei acid hybridization. Several techniques had been in use to anneal RNA molecules to their complementary DNA molecules. In these experiments, both molecules can be in solution, or the DNA can be immobilized in a solid or semisolid matrix, or it can be attached to a nitrocellulose filter. The resulting molecular hybrids are generally detected by scintillation counting of radioactive RNA after treatment with ribonuclease to remove unhybridized RNA. For nucleic acid hybridization in cytological preparations, certain modifications must be made. Here, the DNA molecules are embedded in chromatin containing a variety of protein molecules. The DNA in the chromosomes is native DNA. In solution, one may denature the DNA and immobilize the single-stranded molecules. In the chromosomes, the DNA must be denatured and must be held in the denatured state or there is no hybridization. Finally, the RNA molecules used as a probe must be extremely radioactive because the autoradiographic technique is not as sensitive in detecting radioactivity as scintillation counting.

Since, in a given genome, a particular gene is not represented by numerous copies, the *in situ* hybrid, if formed, will be extremely difficult to detect in autoradiographs. The polytene chromosomes, with each

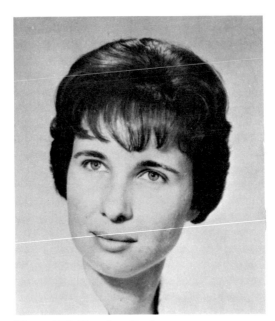

Figure 17.1. Mary Lou Pardue.

gene represented about 1000 times, would be a suitable material. But amphibian oöcytes are even better material. In somatic cells, the ribosomal genes, although mildly reiterated, are not sufficient to be detected in neutral CsCl density gradient as a separate peak. In pachytene and diplotene oöcytes, however, the amount of this gene is dramatically increased (amplified), and it is even possible to isolate the fraction using a preparative ultracentrifuge. At pachytene the oöcyte nuclei are enlarged, and one side of each nucleus is occupied by an enormous mass of amplified copies of ribosomal DNA (more than 1000 times the usual number), known as the *cap*. The amphibian pachytene therefore, is a superior test material for *in situ* hybridization using labeled rRNA as a probe.

As mentioned, the radioactivity of the probing molecule (in this case ^{3}H-rRNA) must be extremely "hot" in order to get sufficient registration in the autoradiographs. Gall and Pardue used a cell culture of the toad *Xenopus* as the source of rRNA. When they added a whopping amount of tritiated uridine to the medium and incubated the cultures for up to seven days, they were able to extract rRNA with approximately 2×10^5 cpm/μg. For molecular biology, this radioactivity is more than sufficient to do many experiments, but for *in situ* hybridi-

Figure 17.2. Joseph G. Gall.

zation, it is barely usable, because the autoradiographs still require months of exposure time.

To prevent sloughing of the cells off the cytological slides, which must go through a number of treatment steps, the slides were first coated with a very thin layer of gelatin in a chrome alum solution. The cellular material was squashed in the usual way for cytological preparations. The slides were treated with a weak HCl solution (thought to remove basic proteins) and then ribonuclease to remove cellular RNA. These two steps were not used in the original report (Gall and Pardue, 1969). The slides were then treated with 0.07 N NaOH to denature the cellular DNA; a two-minute treatment with NaOH at this concentration allows DNA denaturation without causing a loss of the morphological integrity. The labeled RNA molecules dissolved in saline citrate solution (SSC) were layered over the fixed cells and the preparations were incubated (usually overnight in a moist chamber at 65°) in the hope that DNA–RNA hybrid could be formed at the gene loci in question. The slides were treated again with ribonuclease to remove unhybridized RNA molecules, and autoradiographic film was then applied in the usual manner.

Joe and Mary Lou found what they had hoped to find:

heavy labeling over the cap region of the pachytene nuclei in amphibian oöcytes, less heavy in oögonia and leptotene, and very light grain register over the diploid follicle nuclei. The results indicated that the procedure for *in situ* hybridization was well on its way as a useful procedure for mapping DNA in chromosomes. However, a defect had still to be corrected, namely, the insufficient radioactivity of the RNA molecules for efficient autoradiography. To achieve a high specific activity, Joe and Mary Lou did the next logical thing: they synthesized ³H-RNA in a cell-free system from DNA templates. If the template DNA fraction was pure, all the RNA transcripts should be complementary to the DNA sequence. The radioactivity could be increased because one could employ ³H-labeled nucleotides. Joe and Mary Lou used the *Xenopus* oöcyte DNA fraction that contained the amplified ribosomal cistrons. The ribosomal RNA so synthesized reached a specific activity of 7×10^6 cpm/μg, and the exposure time of the autoradiographs could be reduced to a matter of days.

After testing the distribution of the ribosomal RNA in the polytene chromosomes of *Drosophila, Rhynchosciara,* and *Sciara* with her collaborators, Mary Lou started work with Joe to test the cytological distribution of the famous satellite DNA of the mouse. This was a wise choice, since the mouse satellite DNA fraction comprises nearly 10% of the genome, and much was known about its biophysical properties. With the method for synthesizing transcripts in the cell-free system, it would not be too difficult to do the *in situ* hybridization experiments, so long as the satellite DNA fraction could be isolated and purified. The results were startling: The satellite DNA is located at the centromeric areas of every mouse chromosome except the Y chromosome. The chromosome arms other than the centromeric areas were completely devoid of, so far as the *in situ* hybrids could reveal, the satellite DNA sequence. The demarcation between the labeled segments and the unlabeled segments was exceedingly sharp.

In the fall of 1969, Mary Lou presented all these data at the annual meeting of the American Society for Cell Biology. I happened to chair the session, which was packed to capacity. It was one of the most exciting papers I had ever heard. After her presentation, I commented that I wished I had done this piece of beautiful work. My colleague, Frances Arrighi, was also there. She and I decided after the session that we must learn the entire procedure in detail in order to do some *in situ* hybridization work of our own. We cornered Joe and Mary Lou to see if they would let Frances visit their lab at Yale to learn the ropes. Both kindly consented; Frances spent about ten days in Joe's lab in December 1969 and learned the procedure in detail.

The paper of Pardue and Gall (1970) on mouse satellite DNA appeared in the spring. Here I should add that the *in vitro* transcription method is not the only way to obtain the radioactive probing molecules. Kenneth W. Jones (1970) did similar work using ^{3}H-labeled satellite DNA by feeding ^{3}H-thymidine to the mouse cell cultures and performing DNA–DNA hybridization *in situ*. The only problem is that the DNA had a very low specific activity so that Ken's autoradiographs were not as convincing as those presented by Pardue and Gall. In all subsequent works, Ken also used the *in vitro* transcription system.

In the first few months of 1970, Frances and I assembled all the necessary equipment and supplies for doing the *in situ* hybridization experiments. We were still interested in identifying ribosomal cistrons on metaphase chromosomes of various animals. We tried to feed the mammalian cells in culture ^{3}H-uridine to isolate ribosomal RNA, but the cells apparently died of internal radiation. The rRNA showed very poor specific activity (about 5×10^4 cpm/μg) and was useless. We tried to ellicit collaboration from those who knew more about rRNA than we, but got no enthusiastic response. Some molecular biologists did not accept the *in situ* hybridization data at all. One day at the luncheon table we were complaining again, and Grady Saunders said he would be glad to collaborate if we would use his human repetitive DNA fractions as a probe. This casual conversation started our collaboration with Grady for several years.

We used the repetitive DNA fraction from the hydroxyapatite column eluted at Cot = 0–1 as template to

synthesize radioactive complementary RNA (^{3}H-cRNA.)
Our first attempt did not yield cRNA with as high a
specific activity as we wanted, and the pilot autoradio-
graph did not have a heavy grain count. But it was in this
in situ hybrid that we found the differential staining between euchromatin and heterochromatin of human chromosomes (Chapter 18). Anyway, we proceeded to do a
number of experiments on repetitive DNA, both of
human and of mammal origin. In the meantime, many
investigators also applied the technique to study a
number of DNA and RNA species, including 5S RNA,
tRNA, the histone messenger, satellite DNA fractions of
man, chimpanzee, cattle, *Drosophila*, and many others.
But my original dream, to identify the nucleolus organizers in the mammalian species with no secondary constrictions remained unfulfilled.

Actually, several teams of biologists were engaged in
studying the 28 + 18S rRNA, using metaphase chromosomes of animal species such as the rhesus monkey, the
mouse, and human, but the quality of the autoradiographs did not completely satisfy my aesthetic appetite.
For example, the human nucleolus organizers are supposed to be in the short arms of the ten D-group and
G-group chromosomes (Nos. 13, 14, 15, 21, and 22). Each
short arm has three subdivisions: the stubby basal segment, the achromatic secondary constriction (NOR?),
and the terminal segment known as the satellite. *In situ*
hybrid using 18 + 28S rRNA as a probe (Henderson *et
al.*, 1972) showed grains over the short arms but no
definite conclusion could be made regarding the exact
segment.

In 1973 at the International Genetics Congress in
Berkeley, H. J. Evans proudly pulled a photomicrograph
out of his pocket to show it to me. It was an autoradio-
graph of a human metaphase with silver grains localized
only in the secondary constriction regions of the acrocentrics. I gasped and asked John who had done it. He told
me it was a collaborative work between Mary Lou and
his people. The picture was so beautiful that I immediately saw what I had hoped to do. Fortunately, Mary
Lou and I were speakers of the same symposium, so I
found her the next day. There we agreed to work together
on nucleolus organizers of various mammals. We decided
on the species to use, and I would supply her with the
cytological preparations according to her specifications.
She would then perform *in situ* hybridization with ^{3}H-18

+ 28S rRNA extracted from her *Xenopus* cell cultures. We would then both examine the *in situ* hybrids when they were ready. This was one of the most pleasureable collaborations I ever had. We found that in most species the secondary constriction is the NOR, but minor rDNA sites do not always exhibit themselves as secondary constrictions. In species without visible secondary constrictions, the NORs are located at the telomeric ends of some chromosomes or at the centromeric ends of some telocentrics, forming invisible short arms.

In mammals, ribosomal cistrons had never been known to locate in the sex chromosomes, expecially the Y-chromosome. Those found in the compound X chromosomes of the rat kangaroo, the Indian muntjac and the Seba's fruit bat *Carollia Perspicillata* are carried originally by the acrocenric autosome which had been translocated onto the X. One of the species I sent to Mary Lou for *in situ* hybridization was *Carollia castanea* which had a tiny Y chromosome. Mary Lou telephoned to say that she could not find the Y in the autoradiographs, but invariably saw a cluster of silver stains in every metaphase plate. She was wondering whether the grain cluster was over the Y. I told her that the grain cluster must be some background debris because no mammalian species had ribosomal genes on the Y. So we made a wager for a martini. I then carefully examined the slides, took the photographs and degrained the autoradiographs. I lost the bet. I thought I was an observant person, but I had to conclude that the young lady is a superior molecular cytogeneticist. The Y chromosome of *C. castanea* does contain rDNA, a unique situation in mammals.

The *in situ* hybridization technique not only gives cytology a molecular basis (or molecular biology a cellular basis), but also yielded a marvelous byproduct, the technique to identify constitutive heterochromatin. Although it is still premature to say that the resolution can be so sharp as to unequivocally locate single-copy genes (by reverse transcriptase, for example, from polyadenylated RNA to obtain ^{3}H-cDNA), the prospect is hopeful that some day we shall succeed.

18 Junk DNA and Chromatin?

In the early days of molecular biology, investigators used mainly prokaryote DNA for biophysical and biochemical characterization. They found, by using CsCl density-gradient centrifugation to measure buoyant density as a guide for estimating base composition, that the DNA base composition of microorganisms varied from a high adenine–thymine (AT) content to a high guanine–cytosine (GC) content; but within each species, the base composition is constant and homogenous. In neutral CsCl density gradient, the DNA of each species shows a sharp peak in the densitometer tracing.

Eukaryote genomes are naturally much more complex then bacterial or phage genomes, and their DNAs are expected to be more heterogeneous. Thus, in CsCl gradient it was not a surprise to find a broader band. The buoyant density is measured by its peak value, similar to using T_m to measure melting profile. However, it was indeed a surprise when Noboru Sueoka (1961) found that the DNA of the crab (*Cancer borealis*) segregated into two distinct components in CsCl density gradient, a broader major band and a very sharp minor band consisting of nearly pure AT pairs. Noboru was working in Paul Doty's laboratory at Harvard, and he thought it might be interesting to take a look at the DNA of some eukaryotes. No one in that laboratory knew what material he should try first, so they suggested that he go to the Biology Department for consultation. We went there one afternoon and found that the laboratory class had an exercise in dissecting crabs. The laboratory instructor simply gave him some crab tissues from which to extract DNA. It was luck that crab DNA happens to possess such a unique minor band (now known as the satellite DNA), because distinct satellite DNA of this type is not common even in eukaryotes (Sueoka, 1961).

Another lucky situation is that the laboratory mouse (*Mus musculus*) also has a distinct fraction of satellite DNA, independently and almost simultaneously found by Saul Kit (1961) and Waclaw Szybalski (1961). This satellite fraction is not as AT-rich as the crab satellite, but is sufficiently different from the main band that it can be isolated and purified. Subsequent work on many mammals showed that *M. musculus*, with such a distinct AT-rich satellite DNA fraction, is indeed unique. Most mammalian DNAs show no satellite or poor differentiation of satellite fraction(s) in neutral CsCl gradient. If the DNA of other mammalian species had been studied before that of the mouse, genome analyses of higher animals probably would not have advanced as rapidly as it did.

Detailed analyses of the eukaryote satellite DNA fractions revealed a very striking fact: The base sequences within the satellite fractions are usually very short, but they are repeated numerous times. When one sequence with base composition sufficiently different from the bulk DNA is repeated numerous times, a satellite fraction can be detected. In the mouse the satellite represents nearly 10% of the total genome. But when the base composition of the repeated sequence is close to that of the bulk (main band) DNA, the satellite fraction is masked by the main band in ordinary CsCl gradient. As biophysical techniques improved (adding antibiotics with affinity for a special base, adding heavy metal ions to a Cs_2SO_4 gradient, etc.), satellite fractions masked by the main band could also be isolated in any eukaryote organism. In many cases (human, for example) a genome may contain multiple satellites differing from one another in base composition, base sequence, and amount (for details see Flamm, 1972; Saunders, 1974).

Using a different approach, the reassociation kinetics of sheared and denatured DNA, Roy Britten and his associates also found that repetitive DNA sequences are present in eukaryotic genomes, and that the number of repeats may vary from as high as a million copies down to a few copies. In essence, the procedure involves shearing DNA into certain lengths, denaturing the DNA fragments, and measuring the reassociation rates of the fragments as a function of time. The greater the number of repeated copies, the faster would be the reassociation rate. This method for genome analysis is highly complementary to the density-gradient analyses, because satellite fractions always show faster reassociation rates. The mouse satellite, with approximately a million copies per genome, anneals almost instantaneously. Using Britten's method, one may isolate repetitive DNA sequences from hydroxyapatite columns according to their reassociation rates (number of copies).

The question of why the eukaryote genomes are so enormous has been an enigma to biologists. If every base pair is used for genetic information, the genomes simply contain too many genes. The discovery of repetitive sequences, whether tandemly repeated, invertedly repeated, or dispersed, added credence to the notion that not all the DNA represents true genes. But it took the development of the *in situ* nucleic acid hybridization technique (Chapter 17) to demonstrate that the satellite DNA fractions are cytologically clustered in some segments of chromosomes. In the mouse karyotype, the famous satellite DNA is concentrated in the centromeric areas (Pardue and Gall, 1970).

In the summer of 1970, when Frances, Grady, and I were testing the locations of highly repetitive human DNA (the Britten technique), we noted a differential staining along the human metaphase chromosomes in the *in situ* hybrids. The centromeric areas showed darker Giemsa staining than the chromosome arms, and the size of these deeply stained segments appeared to be constant for each chromosome. For example, the long arm of chromosome Nos. 1 and 16 each had a large piece, whereas chromosome No. 2 had a small piece almost directly over the centromere. The most impressive characteristic was the human Y chromosome, which had a heavily stained distal segment in the long arm, corresponding to the brightly fluorescent segment when the human chromosomes were stained with quinacrine. We used the lymphocytes of a colleague in our Biology Department because he has a long Y chromosome. The poor man was bled a number of times by us but he never refused. Incidentally, he later became the department chairman. Of course, I also looked at my own chromosomes. They were of the mundane variety without an impressive Y chromosome. Frances never let anyone take a blood sample from her to examine her own chromosomes.

Frances and I knew we had discovered something significant, probably a method to identify heterochromatin in the human karyotype. Since we did not find a whole chromosome of the C group stained heavily throughout, we surmised that facultative heterochromatin (the repressed X) did not respond to the treatments. Therefore, the darkly stained segment must represent constitutive heterochromatin. Frances told me that Mary Lou had observed a similar structure in the centromeric areas of the mouse chromosomes in her *in situ* hybrids. We suspected that the radioactive complementary RNA used in *in situ* hybridization was not responsible for this differential staining, but the treatment procedure (NaOH denaturation and SSC incubation) was. Therefore, we repeated the experiments a number of times without the complementary RNA, varying the conditions. Differential staining was again noted when the slides were treated with alkaline and incubated in SSC. At that time, a pediatrician in Los Angeles had a triple-Y child and asked us to verify his finding. We gladly applied the procedure to a couple of his slides and found three typical Y chromosomes with heavily stained distal segments.

We decided that this discovery was so important that our report should be published as fast as possible, so we submitted the manuscript to *Lancet*. In about a week we received the letter of rejection, stating that this paper, though interesting, had no medical application. We were reminded by the editor that *Lancet* is a medical journal. We then submitted the paper to *Cytogenetics*, which finally published the report (Arrighi and Hsu, 1971).

To be fair, Frances and I were not the only team to discover this heterochromatin staining technique. T. R. Chen and Frank Ruddle (1971), in the same university as Gall and Pardue, found the same phenomenon when they used the *in situ* hybrid procedure to differentiate mouse chromosomes and human chromosomes in cell hybrids. Yunis and Yasmineh (1971) used heat denaturation and renaturation to treat cytological preparations. They, too, found the differential staining property in several species. The quality of Yunis and Yasmineh's heterochromatin preparations, however, was not as consistent as that induced by alkaline denaturation.

Afterward, Mary Lou told me that she deliberately

downplayed her observation on this differential staining in her report in *Science* with Joe in the hope that not many people would take notice, so that she could work on it when she had time. Actually, prior to writing up our manuscript, I did call Joe to see if he and Mary Lou would object to our writing a paper on this subject as a staining procedure. He said no; so we went ahead. In the meantime, we applied the procedure to a variety of mammalian cells and found it to work in all cases. We were sure by that time that the procedure enabled us to reveal constitutive heterochromatin, because even in metaphase chromosomes of *Drosophila melanogaster* the distribution of the heavily stained segments coincided with the known heterochromatin pattern observed in prophase and prometaphase.

The distribution of constitutive heterochromatin in mammalian species was found to be quite variable. In most cases, heterochromatin is centromeric; but in others, it may be interstitial, terminal, or even comprise total arms. Furthermore, within a species, polymorphism in heterochromatin content is very common. In the human population, for example, practically no two individuals have identical heterochromatin patterns (Craig-Holmes and Shaw, 1971). In the fall of 1971, the Paris Conference for the standardization of human chromosome nomenclature suggested the term C-band for constitutive heterochromatin (Figure 18.1).

As more and more investigators applied the *in situ* hybridization technique to identify the cytological loca-

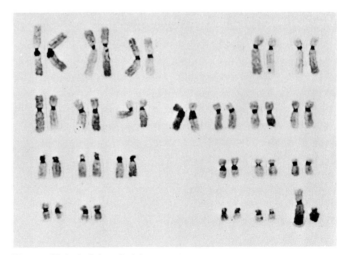

Figure 18.1. A C-banded human karyotype.

tions of repetitive or satellite DNA sequences of various animals (and, later, insects and plants), it became apparent that highly repetitive sequences are the principal components of constitutive heterochromatin. In the mouse, there is only one major satellite DNA, and it is located in the paracentric C-band areas. In the human genome, there are several different satellites. Each of the heterochromatin pieces contains a mixture of these sequences in different proportions. For example, the C-bands of chromosome Nos. 1 and 16 contain a high proportion of Satellite II but that of chromosome 9 contains a high proportion of Satellite III.

It appears that whenever there is a concentration of highly repetitive sequences, a C-band will manifest itself, regardless of base composition. In the mouse genome, the satellite DNA is AT-rich. In the human genome, the satellite fractions encompass a wide spectrum of base composition from highly AT rich to highly GC rich. In cattle, and in the kangaroo rats, all satellites are GC rich. Yet all of them respond to the cytochemical procedure and reveal discrete C-bands. The mechanism for the C-band staining is still unclear. At first, many thought that it may be due to the faster reassociation rate of the repetitive DNA than the unique sequences located in euchromatin. However, several groups of investigators later found that there is a differential rate of DNA loss in these preparations after the treatment procedure with the DNA in the heterochromatin preferentially preserved. Thus, the differential rate of DNA loss may be responsible for the differential staining. However, the reason for the retention of the highly repetitive sequences may still be that annealing rate is faster than the unique sequences.

It is still a puzzle why eukaryote genomes invariably possess a certain amount of highly repetitive DNA (or constitutive heterochromatin). It is also a puzzle why repetitive DNA sequences vary so much in amount, base composition, and sequence. Even closely related species may have entirely different characteristics in their heterochromatin and repetitive DNA. In *Peromyscus*, for example, the amount of constitutive heterochromatin, in the form of C-banded short arms, varies drastically from species to species. Preliminary analyses showed that the satellite DNA sequences are not homologous among species. In the mouse genus *Mus*, most species possess 40 telocentric chromosomes and heterochromatin in the centromeric segments, but *Mus dunni* from India have a variable number of heterochromatic short arms. When the highly repetitive satellite DNA fractions are compared, each species has its own satellites. Nucleic acid hybridization experiments of Sutton and McCallum (1972) showed that the *M. caroli* genome has some degree of homology with the satellite DNA of the laboratory mouse, *M. musculus,* even though the satellite fractions of the two species are not the same. Although no detailed sequencing has been done, the situation in *Mus* may be analogous to that in *Drosophila virilis* studied by Joe Gall and his associates (Gall *et al.*, 1974). In *D. virilis* they found three satellite fractions differing in neutral CsCl buoyant density, but the base sequences of these satellites are, without doubt, related. All of them are heptomers with tandem repeats of seven bases:

I   5`——A–C–A–A–A–C–T——3`
II   5`——A–T–A–A–A–C–T——3`
III 5`——A–T–A–A–A–T–T——3`

Thus, from Sequence I to Sequence II, there is only one base change, and from Sequence II to Sequence III, another base change. It is quite possible that a similar pattern exists in the evolution of mammalian satellite DNA; therefore, partial homology between biophysically dissimilar satellites is not unexpected. But the interesting question is how a mutational event occurring in one sequence could replace the original sequence distributed in many chromosomes and become the predominant one in another species, again distributed in all chromosomes. Thus far, there is no good system to probe this point experimentally. One thing we do know: Repetitive sequences are less conservative than gene sequences, presumably because the repetitive sequences are not informational and changes in them are, therefore, tolerated by the organism.

Similar to the origin of the highly repetitive sequences, the function of these sequences is not understood. To call the repetitive sequence "junk DNA" is perhaps injudicious, but none of the speculated functions of repetitive DNA or heterochromatin (junk chromatin?) has any backing by experimental data. Since no eukaryote organism has been found to be completely devoid of repetitive DNA and heterochromatin, it is inconceivable that nature could have made repeated blunders.

The studies on heterochromatin have come a long way since the early days of Heitz and his contemporaries, but understanding it is still a long way off.

19 The Giemsa Magic

Gustav Giemsa (Figure 19.1) probably never dreamed that his concoction, the Giemsa stain, which was devised to stain malaria in blood preparations, would make a great contribution to cytogenetics. Giemsa was born in 1867 in Blechhammer, and started his career as a pharmacist and chemist in East Africa. In 1900, he was appointed head of the Department of Chemistry of the newly founded Institute of Naval and Tropical Diseases in Hamburg, Germany. He developed chemical insecticides and chemotherapy with derivatives of quinacrine, arsenus, and bismus. His association with malaria research led him to investigate the factors involved in the Romanowsky's stain, which was a mixture of methylene blue and eosine. He found that a degradation product of methylene blue, methylene azure, was responsible for the staining of chromatin and that eosin contributed to it by binding the azure-incorporated chromatin. He then mixed azure and methylene blue in equal amounts to improve the stain and finally used a mixture of azure, eosin, glycerine, and methanol. Staining was also improved by using buffer solutions. Dr. Giemsa never used his name for this stain, and many modern cytogeneticists who use Giemsa stain daily do not know Giemsa is a person's name.

Classic cytologists used acetic orcein, acetic carmine, gentian violet, hematoxylin, and other stains for chromosome preparations, but seldom used Giemsa. Hematologists used Giemsa for blood and bone marrow preparations. The popularity of Giemsa stain among human and mammalian cytogeneticists probably stem-

Figure 19.1. Gustav Giemsa. (Courtesy of Dr. Rudolph Pfeiffer.)

med from Peter Nowell. When Peter was studying human lymphocyte cultures (Chapter 7), he naturally used the Giemsa blood stain. When Paul Moorhead and his associates (including Nowell) described their lymphocyte culture method for human cytogenetic preparations, their recipe called for Giemsa staining. This procedure has been so reproducible and so convenient that practically all human cytogeneticists (and, later, mammalian cytogeneticists) followed.

Giemsa is not superior to orcein for conventional cytological staining, but the magic came when the C-band technique was invented. As mentioned in the last chapter, Frances Arrighi and I submitted the human heterochromatin manuscript to *Lancet* and our paper was rejected. We then rewrote the manuscript and submitted it to *Cytogenetics*. On a business trip to Denver, I told Kurt Hirschhorn and Herb Lubs about this incident, and gave each of them a manuscript copy to try out the technique. Frances and I later sent out more preprints to friends. In the autumn of 1970, a human cytogenetics conference was held at Colorado Springs, and Margery Shaw gave a report for us and distributed more preprint copies.

Apparently, many cytogeneticists tried the procedure, including Maximo Drets who was working in Margery's laboratory as our next-door neighbor. Most human cytogeneticists used air-dried or flame-dried preparations. When the C-band procedure was applied to flame-dried slides, a new set of crossbands emerged. In 1971, several papers were published describing procedures to demonstrate the new Giemsa banding techniques (Sumner *et al.*, 1971; Patil *et al.*, 1971; Drets and Shaw, 1971). At least two were direct derivatives of our C-band procedure.

In the Paris Conference that fall, a careful study was made to compare the quinacrine fluorescent banding (Chapter 16) and this new Giemsa banding (called the G-band) in human chromosomes. The two systems match almost band by band, with the brightly fluorescent Q-bands being equivalent to deeply stained G-bands, and the dull or nonfluorescent Q-bands corresponding to lightly stained or unstained G-bands. Thus, according to the interpretations made in Q-banding (bright fluorescence indicates high adenine–thymine content and dull or no fluorescence indicates high guanine–cytosine content), dark G-bands indicate high AT content and negative G-band indicates high GC content. One of the most striking cases can be found in the chromosomes of cattle, which have four satellite DNA sequences, all GC rich and all located in the C-band regions near the centromeres. In G-band preparations, the centromeric areas of the cattle chromosomes are completely unstained as if these segments have been lost.

Frances and I also noted faint crossbands along human chromosomes in some of our C-band preparations. But we were engrossed in *in situ* hybridization and C-banding, so we decided to let the investigation of the crossbands wait. It was, therefore, poetic justice that we scooped Mary Lou on one project while we were scooped by other investigators in another. At any rate, these earlier procedures were quickly overshadowed by the trypsin treatment technique (Seabright, 1971; Wang and Fedoroff, 1972). I heard of Marina Seabright's trypsin technique from Charlie Ford before her paper was published. We immediately tried it and found the technique marvelous. We have been using the method ever since, although many other agents and procedures were later found to be equally successful in inducing banded chromosomes.

The G-banding technique gave cytogeneticists a more convenient tool than Q-banding for chromosome recognition for these reasons:

1. It does not require UV optics.
2. It is more permanent.
3. It shows many more fine bands than quinacrine when the preparation is well done.

Quinacrine fluorescence fades fast, and it is necessary to photograph immediately without careful observation; in many cases, the fluorescence fades during photographic exposure. The G-bands have no such disadvantage.

Interestingly, although the G-banding technique was originally a derivative of the C-band procedure, C-bands are not always deeply stained in G-band preparations. The cattle C-bands just mentioned are excellent examples. In the human karyotype, the largest C-band blocks are found in chromosome Nos. 1, 9, 16, and Y. Yet in G-banded preparations, the C-band area of chromosome 9 is usually unstained and the Y-chromatin (brightest in fluorescence) is not extremely dark in G-banded slides. Presumably the C-bands indicate repetition of DNA sequences irrespective of the base composition, whereas the Q- and G-bands indicate base composition irrespective of sequence repetition. Thus, a chromosome segment containing highly repetitive sequences rich in GC content will be C-band positive but G-band negative.

Simultaneous with the discovery of G-banding, Dutrillaux and Lejeune (1971) found another method for inducing chromosome crossbanding. They heated the cytological preparations in hot phosphate buffer and then stained them in Giemsa. The bands, interestingly, are the opposite of the G-bands: The darkly stained zones in G-band preparations were lightly stained whereas the lightly stained G-bands were deeply stained. The Paris Conference suggested that these bands be called the R-bands, meaning reverse bands. The Paris Conference also wisely suggested that no band, light or dark, should be called interband, because the intensity of staining can be reversed depending upon the procedure used.

R-banding has its virtues. In the human karyotype, the terminal segments of most chromosomes are G-band light and Q-band dull. Therefore, in some preparations it is difficult to determine the chromosome ends. In R-band preparations, this problem is easily resolved. A modification of the R-banding procedure is to use acridine orange fluorescence after the slides are cooked in the phosphate buffer solution. The brilliant R-bands are something to behold. R-bands can also be induced by treating fixed cells with chromomycin A_3 (specifically binding GC-rich DNA) and later treating the cells with pancreatic DNase (Schweizer, 1977). Apparently chromomycin binding makes the GC-rich DNA less vulnerable to the nuclease digestion than the AT-rich DNA, thus yielding an R-band pattern.

As a routine procedure, most cytogeneticists prefer G-banding. Figure 19.2 presents a G-banded human karyotype. It is not the best ever produced; I use it merely because I made it. Every visitor who comes to our laboratory to learn cytogenetic techniques must set up his or her own blood cultures for chromosome analysis. The one shown in Figure 19.2 was taken from a lymphocyte culture of a young lady who worked with us. Incidentally, from the limited number of visitors, we discovered two cases of abnormal karyotypes.

The banding techniques quickly caused the conventional cytogenetics to become antiquated. A trisomy in the D-group of human karyotype is no longer referred to as D-trisomy, because every chromosome within a group

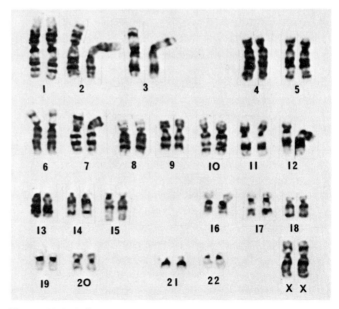

Figure 19.2. A G-banded human karyotype.

can be positively recognized. Deletions, translocations, and inversions can be identified, with the break points located. Many syndromes, determined as cytologically normal during the prebanding era, have been found to have subtle but consistent chromosome abnormalities. And such cases are steadily increasing, especially when less condensed chromosomes (prophase chromosomes) are used for banding.

During recent years, numerous new banding techniques have been published, some of which are slight modifications of the existing procedures. Although they make some technical improvements or confirm the existing banding systems by using different agents, the significance cannot measure up to that of the original G-banding and R-banding, especially G-banding. However, a few special techniques, designed to demonstrate specific areas of chromosomes, are noteworthy. These include the Cd-banding for staining centromeres, the N-banding and silver procedures for staining nucleolus organizer regions, and the technique for differentiating sister chromatids.

Cd-banding.

This wonderful procedure (Eiberg, 1974) has not been fully exploited by cytogeneticists, probably because there is no such need. Giemsa staining following the prescribed procedure will show a deeply stained spot directly over the centromere of each chromatid. It may indeed stain some special proteins associated with centromeres.

Nucleolus Organizer Regions (NOR).

We mentioned (Chapter 17) that the achromatic nucleolus organizer regions represent the sites of ribosomal cistrons, and that in some mammalian species (e.g., the Chinese hamster) the NORs may be telomeric. Most of these conclusions came from *in situ* hybridization experiments using labeled rRNA as a probe. However, *in situ* hybridization has its intrinsic disadvantages. First, it requires expertise in molecular biology and heavy equipment. Second, it is expensive. Third, it is time-consuming. And finally, the resolution is not ideal because the reduced silver grains in the autoradiographs may spill over the neighboring chromosomal regions from the source of radiation.

Not every cytogenetic laboratory can perform such experiments as a routine procedure. It would be highly desirable to find a method to stain NORs. When the achromatic secondary constrictions are conspicuous, one may simply identify the NOR by using cytological preparations stained with conventional nuclear stains. However, when the secondary constrictions are small, ambiguity results. Moreover, some secondary constrictions are not NORs. When the NORs are terminal, it is not possible to demonstrate them because an achromatic segment can be observed only when it is bordered by chromatic segments on both ends. A terminal achromatic segment will invariably escape notice. Thus, Matsui and Sasaki's (1973) N-band technique came at an opportune time. Their method is similar to that of R-banding, but the treatment is more severe. In a follow-up paper, Funaki *et al.* (1975) indeed demonstrated, by N-banding, that the NORs of the Chinese hamster are telomeric.

Unfortunately, many laboratories (including mine) were not able to consistently induce good N-banding. Luckily, an ammoniacal silver staining technique (Howell *et al.*, 1975) for staining NOR just became available, and my colleagues (Goodpasture and Bloom, 1975) applied a modified procedure to stain the chromosomes)f all the species on which other investigators had definitive data from *in situ* hybridization. The silver staining completely confirmed the *in situ* data and was easy to perform. Surprisingly, we found terminal NOR in many mammalian species. After checking the karyotypes of these species, we indeed found no secondary constrictions. Variability in the number and the locations of NORs have become a useful tool to tackle problems in evolution and medical genetics, and silver-stained preparations can be stained with quinacrine to identify both the chromosomes with Q-bands and the NOR simultaneously.

The silver–NOR technique cleared up a controversy. Two research teams (Henderson *et al.*, 1974; Elsevier and Ruddle, 1975) attempted to identify the locations of ribosomal cistrons of the laboratory mouse, using *in situ* hybridization techniques. Both found that there are three pairs of autosomes that carry the ribosomal genes, but their data did not agree with respect to the identity of the chromosomes. The only difference between the two sets of data was the inbred mouse strains employed in these studies. A relatively comprehensive survey (Dev *et al.*, 1977), using the simpler silver–NOR technique, indeed revealed that inbred mouse strains may differ in terms of

NOR locations. This suggests that synapsis and crossing over between ribosomal sequences situated in nonhomologous chromosomes occur frequently, and, as a result, these genes can be transferred from one chromosome to another. Inbreeding may have segregated and fixed these variants. This type of genetic behavior is probably a common occurrence in karyotypes with multiple NORs, as human cytogeneticists have repeatedly discovered anomalies in the short arms of the D and G group chromosomes.

Sister Chromatid Differential Staining.

I mentioned (Chapter 16) that Sam Latt (1973) found a way to differentiate sister chromatids by 33258 Hoechst fluorescence. To demonstrate such a differential, the cells must be grown in a medium containing 5-bromodeoxyuridine (BUdR or BrdU) for two DNA-synthetic cycles. The cells will incorporate BUdR into their DNA as a thymidine substitute. After two cell cycles, one chromatid will contain DNA molecules that have both strands substituted, and the other chromatid, with only one strand substituted. Apparently the double-substituted DNA quenches the 33258 fluorescence, resulting in one bright and one dark chromatid for each metaphase chromosome.

In the late 1950s, Herbert Taylor (1958) found in his autoradiographs of cells previously pulse-labeled with ^{3}H-thymidine, chromosomes with one labeled and one unlabeled chromatids when the cells were allowed to enter the second mitosis after the removal of labeled thymidine. This was one of the evidences supporting the unineme hypothesis of chromosome structure. He also discovered that exchanges between sister chromatids may occur: one chromatid carries label to a certain point and the label is switched to the sister. This phenomenon is known as the sister chromatid exchange (SCE). Now with the BUdR incorporation method, Latt found clear demonstration of SCE in his materials.

Actually, Zakharov and Egolina (1972) first discovered that chromosome segments with BUdR incorporation showed lighter Giemsa staining than those without BUdR incorporation. This phenomenon was later expanded into several procedures to detect sister chromatid differential and SCE by different research teams, all using Giemsa staining instead of fluorescence. Although the mechanism for SCE is still obscure, SCE has recently been utilized extensively to study the effects of environmental mutagens and persons with genetic defects in DNA repair processes.

The magic of the Giemsa staining in cytogenetics is probably just the beginning, but it must be borne in mind that the banding of metaphase chromosomes, when compared to the fine resolution that can be achieved by the use of the polytene chromosomes of dipterous insects, are still crude. Nevertheless, the contributions of these banding techniques to genetics have already been immeasurable. Cytogenetics has been firmly embedded in many branches of medicine; and the recent amniocentesis for the determination of the prenatal genetic makeup of fetuses has caused controversy in legal, ethical, and religious realms. The applications of cytogenetics to academic problems as well as to human welfare are ever-expanding after the adoption of the banding techniques. One can expect that more refined systems will come into being in the future, and cytogenetics will contribute more to fundamental biology, medicine, and the human society.

20 Parasexual Reproduction

Cell fusion normally occurs *in vivo*. The multinucleated muscle cells and osteoclasts are the products of cell fusion. Fusion of cells in culture was observed by Warren H. Lewis in the 1920s, however, utilization of the cell fusion phenomenon as a tool to study a number of biological problems started with Georges Barski and his associates (Barski *et al.*, 1960). In his time-lapse motion pictures of cells in culture, Barski observed that two cells may fuse occasionally to form a single cell. To prove that fused cells can perpetuate themselves and contain two sets of genomes, Barski used as his test materials two mouse cell lines originally established by Katherine Sanford. These two clone lines were isolated from a mouse cell culture and became transplantable. They differed in their ability to induce tumors when injected into histocompatible mice and in chromosome characteristics. One line had a stemline chromosome number of 55 with no metacentric marker chromosomes, whereas the second line had a stemline number of 62 with from 9 to 19 biarmed markers.

By studying the chromosome characteristics, Barski and co-workers (1961) proved that cells of different lines may fuse to form heterokaryons or "hybrid cells" in mixed cultures. In unmixed cultures, polyploid (double-stemline) cells were always present. But these cells had all the chromosomes doubled, including the marker chromosomes. In the cell hybrids, however, the chromosome number equaled the sum of the two lines (115–116), but the marker chromosomes remained single (9–15).

Presumably, two cells first fuse into a binucleated syncyte, and both nuclei may enter mitosis simultaneously. After this first cell division, the two genomes are mixed and the daughter nuclei each contain two genomes.

Confirmation of the heterokaryon formation phenomenon came from Sorieul and Ephrussi (1961). Boris Ephrussi, a renowned geneticist, immediately saw the possibility of utilizing the somatic cell hybrids to study cell physiology and cell function. Indeed, he made many significant contributions to this area of research in the early period of cell hybridization work. Leo Sachs also seized the opportunity to analyze cell function with this approach. The report of Gershon and Sachs (1963) not only confirmed Barski's observation but also presented additional evidence using, instead of chromosome characteristics, antigenic properties to prove that the hybrid cells contained gene products of both genomes. In this study, these investigators used histocompatibility antigens as markers and two cell lines capable of producing tumors only in the strain of mice from which the cells originated. When cells of one cell line were inoculated into the mice of the other strain, the transplant was rejected. The hybrid cell line did not survive in either mouse strain but it did in the hybrid mice (F_1) from crosses between the two strains. These results indicate that the hybrid cells contained surface antigens of both parental cells. Therefore, they were incompatible with both parents, but became compatible with the F_1 hosts.

These early results suggested that somatic cell hybridization may be a useful tool to study a variety of biological problems. However, the frequency of spontaneous cell fusion to form cell hybrids is exceedingly low. Therefore, developing methods for increasing the fusion frequency became important. Two systems were quickly developed, the selective medium method and the virus method.

The purine salvage pathway depends on the enzyme hypoxanthine guanine phosphoribosyl transferase (HGPRT). When this enzyme is absent (the gene coding for HGPRT is missing or mutated), the cell can rely solely on its endogenous pathway for purine deoxyribonucleotide (hence DNA) synthesis. And when the endogenous pathway is poisoned by such chemicals as aminopterin (a folic acid analog), the HGPRT$^-$ mutant cells cannot survive because DNA synthesis stops.

Similarly, the pyrimidine salvage pathway depends on

the enzyme thymidine kinase (TK). Again, a mutant cell with no TK activity (TK⁻) relies completely on the endogenous pathway for pyrimidine deoxyribonucleotide synthesis: If this pathway is blocked by aminopterin, TK⁻ cells succumb. John Littlefield (1964) placed two mutant cells (HGPRT⁻ and TK⁻) in the same culture vessel and fed them with a medium containing hypoxanthine, aminopterin, and thymidine (HAT), which was devised by Szybalski *et al.* (1962). Both types of cells are expected to die because neither can complete DNA synthesis even though the affected pathways are different. However, if cell fusion occurs, the resulting heterokaryon will survive since the HGPRT⁻ mutant can supply TK and TK⁻ mutant can supply HGPRT. Thus, only hybrid cells can develop colonies in the HAT medium. This method does not increase the frequency of cell fusion, but it eliminates all the parental cells and makes the recognition and selection of hybrids effortless.

In the late 1950s and early 1960s, Y. Okada (1958, 1962) reported that a myxovirus, called HVJ or Sendai (now most biologists use the name Sendai virus), can cause fusion of Ehrlich tumor cells, resulting in multinucleated giant cells. Later Harris and Watkins (1965) found that UV-inactivated Sendai virus can cause cell fusion as efficiently as live virus, thus drastically reducing the danger of viral infection in the research area, particularly animal colonies. Inactivated Sendai virus is still the standard "fusogen" today, even though some chemical fusogens, such as polyethylene glycol, are gaining popularity.

Obviously, the use of Sendai virus for somatic cell hybridization is one of the most significant advances in this area of research, and the horizon expanded when several teams of investigators demonstrated that cells of different origins, including different species, may fuse to form cell hybrids (Harris and Watkins, 1965; Okada and Murayama, 1965; Ephrussi and Weiss, 1965; Yerganian and Nell, 1966). Now one can produce hybrids parasexually between cells of any two species, and the partners are not limited to mammals. Weiss and Ephrussi (1966) showed that in proliferating somatic cell hybrids, both parental genomes (e.g., man/mouse) may be expressed. These discoveries led to a tremendous research activity in, and consequently increased our understanding of, somatic cell genetics.

But in the field of cytogenetics, the findings of Weiss

and Green (1967) should be considered most important. When Boris Ephrussi left the United States to return to Paris, Mary Weiss did not at first move with Boris, but spent some time in Howard Green's laboratory. It was during this period that Mary observed a remarkable phenomenon in perpetuating cell hybrids derived from fusion of cells of two different species. In the man/mouse hybrid cell lines, she found that the human chromosomes were selectively eliminated in the hybrid cells. Previously, the presence of chromosomes of both parental complements was a decisive proof for the nature of the cell hybrids, but now one set of chromosomes could be preferentially eliminated as the hybrid cell lines continued to proliferate. In the report of Weiss and Green (1967), all except 2–15 human chromosomes were eliminated from the man/mouse hybrid cell lines during the 20 cell generations after fusion, but after this time, the karyotypes stabilized at 1–3 human chromosomes. The elimination of chromosomes of one genome showed considerable variability in terms of speed, the number and types of chromosomes retained, and other parameters. When normal human cells fuse with heteroploid mouse cells, usually the human chromosomes are preferentially lost. On the other hand, when normal mouse cells are fused with heteroploid human cells, the reverse is the case, that is, the mouse chromosomes are preferentially deleted from the hybrid genome.

This unexpected phenomenon soon became a powerful tool to map genes on chromosomes. Since many genes determine specific products (proteins) that can be identified, the presence of specific human enzymes and the presence of specific human chromosomes could be correlated in the hybrid cells after chromosome elimination. Take thymidine kinase as one of the easiest examples. If the mouse cell is TK$^-$, its survival in the HAT medium must rely on the TK gene from the human genome. When the human chromosome carrying the TK gene is eliminated in the hybrid, the cell dies. Consequently, at least one human chromosome, the one with the TK gene, must be retained for the hybrid to survive. If a number of clones with a variable number of human chromosomes are examined cytogenetically, one chromosome must be present in all cases and this chromosome must be the one with the TK gene. In the human genome, chromosome 17 was found to carry the TK gene, and the X chromosome was found to carry the HGPRT gene.

This principle can be extended to other enzymes, because many human enzymes differ in electrophoretic mobility from the corresponding mouse enzymes. It is possible to establish whether both genomes or whether specific human enzymes are lost. This information can be correlated with the chromosome data. The presence or absence of a human enzyme should coincide with the presence or absence of a human chromosome. Without the recent banding techniques, however, no precise identification of chromosomes can be made. Fortunately, the development of chromosome banding techniques and the development of gene mapping techniques were nearly simultaneous, so that this new branch of cytogenetics grew rapidly.

Linkage tests can be done using the same principle. When two human enzymes are always present or absent together in man/mouse hybrid clones after extensive human chromosome elimination, the two genes coding for the two enzymes must be located in the same chromosome. Using these methods, many human genes have been unequivocally assigned to individual human chromosomes; these genes could not have been analyzed with the classic human genetic method. Needless to say, many investigators, notably Frank Ruddle, O. J. Miller, D. Bootsma, M. Siniscalco, F. T. Kao, and their associates, contributed a great deal of time and effort to the success of this field.

Many groups of investigators have found that the phenomenon of differential chromosome elimination is not limited to man/mouse cell hybrids. The same methodology can be applied to construct genetic maps of other mammalian species. For comparative genetics, anthropology, and somatic cell genetics, such investigations should bear fruitful information in the future.

One of the applications of the cell hybrid method is the development of plant cell hybridization. Actually, induction of fusion between plant protoplasts was reported earlier than that of the mammalian cell systems (Michel, 1937), but this discovery was ignored until the studies on mammalian heterokaryons were well established (Power *et al.*, 1970). Later, Peter Carlson and his associates (Carlson *et al.*, 1972) produced an interspecific hybrid plant from fused protoplasts. The possibilities of studying plant genetics, plant physiology, and above all, improvements in agriculture and horticulture, seem unlimited.

To cytogeneticists, a byproduct of the parasexual reproduction is just as exciting as other phenomena found in somatic cell hybridization, namely, a method of visualizing the interphase chromosomes. Because of its importance, I describe this system in more detail in the next chapter.

21 Interphase Chromosomes

Potu N. Rao did his graduate work on tobacco cytogenetics. In the 1950s and 1960s, job opportunities for plant cytogeneticists were extremely scarce, and many young botanical cytogeneticists turned to human cytogenetic, tissue culture, and related fields for survival and for new adventure. Potu was one of them. He joined Joseph Engelberg's laboratory at the University of Kentucky, working on cell culture and cell physiology, using HeLa cells as the principal material. During these years, he developed expertise in cell synchronization, and discovered that nitrous oxide is a mitotic arrestant. HeLa cells, when blocked by colchicine at metaphase, have difficulty entering anaphase after the removal of colchicine from the culture medium, but they would divide normally after arrest by nitrous oxide. Rao also perfected the synchronization technique with an excess amount of thymidine to block the cells at the G_1–S border.

In 1968, Potu moved to Denver to join T. T. Puck. He left Lexington mainly because his wife developed severe allergic symptoms and could suffer no more. Engelberg had worked with Puck previously, and recommended Rao. In the fall of 1968, Robert T. Johnson received a Damon Runyon postdoctoral fellowship to work in Puck's laboratory also. Bob received training in Henry Harris's lab in Great Britain, so he was well versed in cell fusion work. Originally, he intended to work on complementation of biochemical mutants following cell fusion.

There were quite a number of young scientists in

Puck's department, and usually all of them went to enjoy a coffee break together. During coffee breaks, they talked about their work and became better acquainted with one another. It was at these coffee breaks that Johnson and Rao exchanged information and decided on a collaborative research in combining cell synchrony with cell fusion techniques to find out what would happen to the cell physiology if two cells in different phases of the cell cycle were fused in a common cytoplasm. For example, one could ask what would happen to DNA synthesis when a G_1 cell fuses with an S cell? Would the G_1 nucleus be stimulated to synthesize DNA prematurely or would the S nucleus be inhibited from continuing DNA synthesis by the G_1 nucleus? Similar questions could be asked regarding fusion of S and G_2 cells and G_1 and G_2 cells.

I must digress for a moment to mention some work of Warren Nichols and his collaborators (1965) on virus-infected cells—particularly measles virus. In the virus-infected cell cultures, these investigators found a small proportion of metaphase cells with, in addition to typical metaphase chromosomes, structures that resembled chromosomes but were much thinner and less intensely stained than the metaphase chromosomes. Furthermore, they appeared to have been shattered into countless pieces. Warren called them *pulverized chromosomes*. If the cultures were labeled with radioactive thymidine, the pulverized material was found to be labeled, whereas the metaphase chromosomes were not. Avery Sandberg also found such a phenomenon in his materials.

Interestingly, the pulverized chromosomes were invariably associated with a metaphase. They were found in virus-infected cultures, but no good explanation could be offered regarding their origin. The mystery was finally solved unexpectedly by Johnson and Rao in their cell fusion experiments.

Fusing cells of different phases of the cell cycle was easier said than done. Large quantities of cells must be propagated and synchronized at various phases so that populations with different phases could be ready simultaneously. After fusion, samples were taken every hour for many hours. Thus, most experiments required an overnight stay in the laboratory.

In one experiment fusing G_2 and S cells, with the S cells labeled, Johnson and Rao wanted to see how the two nuclei moved into mitosis. It was in this set of experiments that they found the pulverized chromosomes in

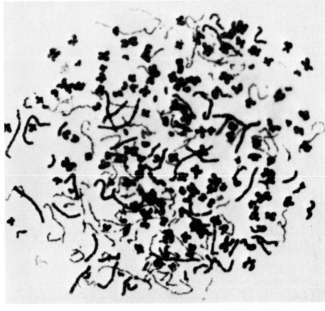

a

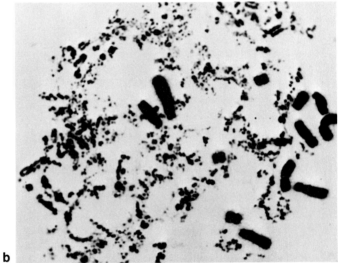

b

Figure 21.1. Prematurely condensed chromosomes: **a,** a syncyte
with fusion between three Hela cells, one metaphase, one G_1 cell
and one G_2 cell. (Courtesy of Dr. W. Unakul.) **b,** A syncyte between
two Chinese hamster cells (CHO) showing a metaphase and a mid-S
cell with "pulverized chromosomes." (Courtesy of Y.-F. Lau.)

large quantities. Invariably, the cell contained a mitotic figure without label and the pulverized portion with label. This indicated that the mitotic figure must have been from the cell in G_2 phase prior to fusion, and the pulverized material, from the cell in the S phase. However, the phenomenon was observed only in the samples when the G_2 nuclei entered mitosis, not immediately after fusion. Therefore, Bob and Potu surmised that the pulverization of the S phase nuclei must have been induced by the mitotic component in the same cytoplasmic envelope of the heterokaryon.

To test this hypothesis, Johnson and Rao designed an experiment to fuse directly mitotic cells and nonsynchronized interphase cells that were prelabeled with ^{3}H-thymidine. Here they found evidence for their hypothesis, that is, the pulverized chromosomes represented the S phase chromosomes that were somehow induced to express themselves when fusing with a mitotic cell. In addition, they found a bonus from this experiment. In the syncytes containing a mitotic figure, there were two more types of chromosome configuration, one with thin, long chromosomes—each apparently consisting of a single chromatid—and the other with thin, long chromosomes—each consisting of two chromatids.

It became obvious to Bob and Potu that they were looking at the interpahse chromosomes of different phases of the cell cycle. The one with single chromatids represented the G_1 cells; the one with pulverized chromosomes, the S cells, and the one with double chromatids, the G_2 cells (Figure 21.1). Apparently, when a mitotic cell fused with an interphase cell, the former induced the interphase nucelus to enter mitosis and to condense its chromosomes prematurely. This marked the first time that biologists, by an experimental procedure, could visualize interphase chromosomes of ordinary somatic cells (Johnson and Rao, 1970).

Duly jubilant, Johnson and Rao rushed to tell Ted Puck of their new findings. Puck was also excited. They decided to call the phenomenon *prematurely condensed chromosomes* or PCC. The pulverized chromosomes are, in reality, S phase chromosomes so extended in the process of DNA synthesis. The heavier chromatin fragments represent the chromosome segments that have completed DNA replication prior to the time of PCC induction, and these fragments are actually connected by microscopically invisible or barely visible chromatin fibers that are replicating. This was later well demonstrated in autoradiographs by Röhme (1975). At any rate, the pulverized chromosomes of Nichols, Sandberg, and other cytogeneticists could now easily be explained as S–PCC in fused metaphase–interphase cells induced by viral infection.

Johnson and Rao proceeded to conduct many more experiments not only to confirm their initial observations but also to expand the horizon to fuse metaphase cells with differentiated (G_0) cells. The chromosomes of the G_0 cells appear to be invariably of the G_1 type, because the PCC always reveal themselves with a single chromatid each. The most striking finding was the induction of PCC of horse sperms, with 30-odd tiny, thin chromosomes.

To be able to observe interphase chromosomes microscopically is certainly a cytological triumph, and the technique should prove useful in studying many problems relating to chromosome structure and chromosome physiology. The elongated G_2–PCC should be excellent materials for more refined subdivisions of chromosomes than metaphase Q-banding or G-banding. One can also analyze the immediate effects of mutagens on chromosomes without waiting for the damaged cells to enter mitosis. Bob did a considerable amount of work on radiation-induced chromosome aberrations, and Potu and his associates, on chemical mutagens. There are many possible applications, but cytogeneticists have not really begun to exploit this technique fully. One reason is probably technical difficulty in synchronizing mitosis, obtaining good Sendai virus samples that give high fusion efficiency, and others. Perhaps the recent polyethylene glycol method, when properly worked out for PCC induction, would stimulate more activities in investigations of interphase chromosomes.

22 Cancer Chromosomes

I have mentioned previously (Chapter 5) that the improvements in karyological techniques (hypotonicity, colchicine, etc.) also stimulated interest in analyzing chromosome constitutions of malignancies. Unfortunately, the lack of criteria for recognizing individual somatic chromosomes still hindered its progress. The general conclusion at that time was that most cancer cells are aneuploid, with or without rearrangements. Of course, only those chromosomes with such morphologic distinction (the marker chromosomes) could be identified as rearrangement products. Many subtle rearrangements escaped notice.

The main purpose of studying the chromosomes of cancer cells was to test the Boveri hypothesis—an imbalance in chromosome constitution causes cancer. Indeed, most cancer cells were found to be karyologically abnormal, but there were no consistent patterns of abnormality in tumors of the same pathological origin. In fact, even in the same animal or in the same patient, the chromosome composition may change considerably as the tumor progresses. Therefore, it became rather difficult to attribute chromosome changes as being the etiological factor.

However, even during the prebanding period, the Philadelphia chromosome (Ph[1]), first reported by Nowell and Hungerford (1960) in chronic myelogenous leukemias (CML), suggested a relationship between specific chromosome changes and specific tumors. One of the G-group chromosomes was distinctly shorter than the others, which appeared normal. Literally thousands of cases

confirmed Nowell and Hungerford's finding. The fact that
Ph^1 in CML is so convincing and consistent is due to two
factors:

1. Ph^1 is the only cytological abnormality in most CML
   cases
2. Ph^1 is so conspicuously small that it is difficult not to
   notice

Since human chromosome Nos. 21 and 22 are morphologically indistinguishable without banding, the Ph^1
chromosome was then regarded as a terminal deletion of
No. 21. This abnormal chromosome is found in all
metaphase figures of bone marrow of CML patients, but
not in lymphocytic organs (spleen, lymph nodes). In
peripheral blood samples, therefore, a mixture of cells,
those with Ph^1 and those without Ph^1, is observed. This
marker chromosome supports a hypothesis in hematol-
ogy, namely, the myelocytic series and the erythrocytic
series share a common ancestor, presumably hemocyto-
blast; but the lymphocytic series is apparently differentiated much earlier during embryonic development. In
lymphocyte cultures without phytohemagglutinin stimulation, the only cells that will enter mitosis are the
leukemic elements. Therefore, all metaphases will show
the Ph^1 chromosome. In cultures with phytohemagglutinin, both leukemic cells from the bone marrow and lymphocytes from lymphatic organs will enter mitosis, thus
yielding a mixture of cell types.

It should be emphasized that not all cases of CML
showed the Ph^1 chromosome. Interestingly, a minority of
cases of CML without Ph^1 (about 20%) are clinically different from those with Ph^1 (Nowell, 1974;
Wang-Peng et al., 1968). Clinicians now use the presence
of Ph^1 as a criterion for the classification of CML. It
therefore appears that, cytologically and clinically, two
types of CML can be recognized, although the two are
pathologically indistinguishable.

Many attempts were made to analyze the cytogenetic
characteristics of other tumors of both human and experimental animals, but the results were inconclusive.
The chromosomal variability among individual tumors of
the same histological type was so extensive that little
sense could be made of the confusing pictures, even
though suggestions that the chromosome changes in
tumors are nonrandom existed in some cases.

Chromosome banding techniques gave the same preci-

·sion tools to cancer karyology as they did to other fields of cytogenetics. Analyses of aneuploidy with respect to the chromosomes involved became feasible, and rearrangements, when not extensive, could be identified. Janet Rowley (1973) first discovered that the Ph[1] chromosome of CML was not a terminal deletion of No. 21, but a translocation between a No. 22 and a No. 9. Approximately half the long arm of a 22, including the middle Q-band, was moved to the terminal position of the long arm of a 9, and presumably only a small piece of the terminal chromatin of 9 was reciprocated (Figure 22.1). This asymmetric translocation, even though it involved a long segment of No. 22, could not be detected by standard karyotyping, because there are 13 or 14 chromosomes in Group C, to which No. 9 belongs. In constructing karyotypes without banding, one simply "forces" the chromosomes to pair.

It is now well documented that all cases of CML with the Ph[1] chromosome contain a translocation involving No. 22. The recipient chromosome, however, is not limited to No. 9. Chromosome Nos. 2, 6, 7, 9, 11, 13, 16, 17, 19, 21, and the homologous 22 have all been found to receive the terminal segment of No. 22. It is also of interest to note that the cytological picture of CML shows additional chromosome aberrations during blastic crisis, and these additional aberrations are again nonrandom, including double Ph[1], trisomy of 8, and trisomy of the long arm of 17.

All of us were taught by our genetics teachers and genetics literature that chromosome breaks occur at random. Experimental data collected from radiation-induced chromosome aberrations substantiated this conclusion. Geneticists used naturally occurring chromosome changes (inversions, translocations) as tags to trace evolutionary trends and to analyze population structure

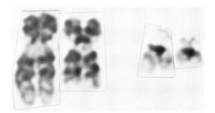

Figure 22.1. Chromosomes 9 and 22 in human chronic myelogenous leukemia, showing the translocation of the distal segment of a 22 (Ph[1]) to the tip of long arm in a 9. (Courtesy Dr. Jacqueline Wang-Peng.)

based on the concept that the probability for the same breaks at precisely the same loci is negligibly small. Therefore, when two inversions with the same breakpoints are found, one can assume that they came from the same ancestor. Probably this concept is still largely correct in natural populations, but recent information on chemical and viral-induced chromosome breakage has indicated that the distribution of chromosome breaks may be nonrandom. The CML cases illustrate that specific translocations between two chromosomes can occur independently and repeatedly. The most recent compilation of data made by Janet Rowley showed that in over 600 cases more than 90% involved the 9/22 translocation, and that the distal portion of No. 22 was invariably attached to the distal end of No. 9q.

Although many investigators around the world made contributions to the field of cancer cytogenetics, the highest concentration of activity has been in Sweden, especially at the University of Lund. Investigators such as Albert Levan, Joachim Mark, Felix Mitelman, Göran Levan, and, in Stockholm, Lore Zech, have done painstaking analyses on a variety of tumors in both human and experimental animals.

The nonrandom nature of chromosome change in Burkitt's lymphomas was described by Manolov and Manolova (1972). In 10 of 12 cases, an extra terminal band in a No. 14 chromosome was found. Zech *et al.* (1976) confirmed this conclusion but added that one of the No. 8s lacked the terminal band. Thus, a specific translocation (between Nos. 8 and 14) seems to be the chromosome change in this neoplasm. Again, the phenomenon emphasizes that specific rearrangements can occur repeatedly.

The human meningiomas and retinoblastomas added other groups of neoplasms that showed nonrandom chromosome changes. In meningiomas a No. 22 was lost in 26 of 27 cases, and monosomy 8 was found in 16 cases (cf. Mark, 1977; Levan and Mitelman, 1975). In retinoblastomas, a No. 13 was found to have an interstitial deletion (Wilson *et al.*, 1973).

In experimental animals, especially rats, a good collection of data is available. The works of Göran Levan, Mitelman, and many Japanese cytogeneticists on sarcomas induced by carcinogens and by oncogenic viruses have also suggested a nonrandom nature of chromosome aberrations (Levan and Levan, 1975). In the mouse, trisomy of chromosome 15 was found to be the specific anomaly in the leukemias of the AKR strain (Dofuku *et al.*, 1975).

The conclusion that chromosome changes in neoplasia are nonrandom (thus suggesting an etiological relationship) was made mainly in tumors displaying a minimum degree of karyological deviations. The vast majority of human carcinomas and sarcomas have not been investigated fully because the cytological pictures are so complex that they almost defy analyses, even with the banding techniques. Several teams of investigators, notably those of Avery Sandberg, George Moore, and N. B. Atkin, have done a considerable amount of work in this area, but the results have not yet been able to support a conclusion. The approach taken by Cruciger *et al.* (1976) may be one way to give this problem a reasonable start. These investigators analyzed seven human mammary carcinomas with such diversified stemline numbers as 40, 43, 43, 49, 57, 59, and 66. Each had a variety of marker chromosomes indicating extensive chromosome rearrangements. Instead of trying to decipher every chromosome, they attempted to determine whether one or more changes were shared by all of these tumors. Indeed, they discovered one such common denominator, a marker chromosome invariably involving the long arm of chromosome 1 (or 1q). It was fortunate that they had several tumors with low stemline numbers and less complex rearrangement patterns. The involvement of 1q could be positively identified in these cases, and this information was used to analyze the cytologically more complex cases. In all seven cases, 1q was involved in a translocation, but the recipient chromosomes were not always the same.

The breast tumor marker chromosome was reminiscent of the Philadelphia chromosome of CML, in which the long arm of a chromosome 22 is always involved but the recipient chromosomes are variable. However, in most CML cases, the translocation involving No. 22 is the only rearrangement. In breast cancers, it is not yet certain whether there are additional marker chromosomes that are shared by most of these tumors. Moreover, seven cases do not constitute a sufficiently large sample for a definite conclusion. The report of Criciger and co-workers only indicates that to look for specific chromosome changes in cancer tissues, one should bear in mind that the morphology of the marker chromosomes may not always be the same. Furthermore,

it points out that isolated cases, when competently
analyzed, should be placed on record because they will
contribute to the information pool as cases accumulate.
Cruciger and co-workers examined G-band karyotypes of
two breast tumors published by other investigators and
again identified the 1q involvement in these.

One of the difficulties in the study of chromosome
constitutions of neoplasia is that tumor cells *in vivo*,
despite their uncontrolled growth capacity, are generally
deficient in mitotic figures for critical cytogenetic studies.
Thus, most of the chromosome investigations were made
on cell lines that may or may not reflect the original
chromosome constitutions of the neoplasms in question.
An alternative approach is to analyze cell clones transformed *in vitro* by oncogenic agents such as oncogenic
viruses or carcinogens. One can procure a relatively large
number of clones and prepare cytological slides as soon
as a sufficient number of cells are available for each.

Investigators in Israel (Leo Sachs and others), Sweden
(Levan and others), and the United States (Joseph Di-
Paolo and others) have done a considerable amount of
work in the cytogenetics of transformed cells. In general,
the following conclusion appears valid: The chromosome
changes in transformed cell lines are nonrandom. Di-
Paolo and co-workers (1973) found, in transformed Syrian hamster fetal cells, a high percentage of clones displaying a monosomic No. 15 (DiPaolo and Popescu,
1976). DiPaolo, however, did not consider this finding
significant because not 100% of the transformed clones
showed the same anomaly. Personally, I think Joe was
asking for an impossible phenomenon. The original cultures were those of fetal tissues. There was no cloning
prior to the transformation experiments. Therefore, the
cultures might contain dozens of cell types in various
stages of differentiation. From the limited data on human
neoplasms, we learned that a specific chromosome
change is associated with a particular type of tumor
(CML, Burkitt's lymphoma, meningioma, etc.). We
would then expect to find dozens of types of cytogenetic
characteristics if all types of embryonic cells are equally
responsive to oncogenic agents. The fact that a high
percentage of the transformed clones showed a common
anomaly suggests that only a few types of the embryonic
cells responded to the transforming agents as target cells.

At the present time, no direct evidence is available to
demonstrate that a specific chromosome change is the

cellular event that initiates neoplasia. How a change in the chromosome constitution of a cell can initiate uncontrolled growth remains speculative; but as other fields of biology, such as genetics and developmental biology, advance, it is anticipated that cancer chromosome research will also advance.

23 Chromosomes and Mammalian Phylogeny

Morphologically, the Indian muntjac and Chinese muntjac are very much alike, but their chromosomes are drastically dissimilar (Chapter 11). The chromosomal differences emphatically suggest that the two species are so completely isolated that no fertile hybrid can be expected even if the two species had an opportunity to produce offspring. However, if the karyotypes of related forms are always as vastly different as the two muntjacs, then cytogenetic characteristics are useless phenotypes for phylogenetic studies because not much information can be extracted in tracing relationships. Fortunately, such instances are not common. In fact, in a number of cases, related species possess identical or nearly identical chromosome constitutions. For example, the karyotypes of many vole species belonging to genus *Clethrionomys,* which inhabit both the Eurasian and the North American continents, are indistinguishable. Similarly, most cat species possess the same karyotype. Therefore, when a traceable change in the karyotype occurs, cytogeneticists can then use the chromosome phenotype as an additional set of criteria to interpret phylogeny. Most cats have a diploid number of 38, but the ocelot has a diploid number of 36. The karyotypes clearly indicate that a centric fusion has occurred between two pairs of acrocentrics of the cat karyotype to form a pair of metacentrics of the ocelot karyotype.

As mentioned previously, Robert Matthey started cytological analyses of vertebrates as an approach to phylogenetic study long before the introduction of newer techniques. However, real progress was made in the 1960s, when several enthusiastic young mammalogists entered the field. A large number of species were cytologically examined and many interesting facts were uncovered. In many instances, cytogenetic data support the conclusions made by taxonomists. For example, the classic classification system of the North American rodent genus *Peromyscus* contained seven subgenera. One subgenus, *Ochrotomys,* has a single species, the golden mouse, *Peromyscus nuttalli,* Hooper and Musser (1964) suggested that this subgenus should be elevated to the generic status (genus *Ochrotomys*) because the morphological and anatomical features are sufficiently different from those of other subgenera. Chromosome analyses made by Patton and Hsu (1967) revealed that all *Peromyscus* species have a diploid number of 48 but *nuttalli* has a diploid number of 52.

In other cases, cytogenetic research definitely revised taxonomy and, therefore, made contributions to systematics. When no distinct morphologic differences exist, taxonomists are obliged to regard similar specimens under the same species. But cytogenetic studies have shown that animals with similar morphology may be entirely different in chromosomal characteristics. In their survey of the karyological characteristics of the cotton rats (genus *Sigmodon*), Zimmerman and Lee (1968) found that *S. hispidus* has a diploid number of 52 in all Southeastern and South Central United States, but in Arizona the animals has a diploid number of 22. Such a difference, analogous to that between the two muntjacs, suggests that the two forms of *S. hispidus* are genetically so distant from each other that they could not possibly be one species despite their morphological similarity. The Arizonan form has since been recognized as a distinct species, *S. arizonensis*.

In Madras, India, Robert Matthey obtained some 30 specimens of mice suspected to be *Mus booduga*. However, chromosome analyses showed that there were two distinct types. One of them had 40 telocentric chromosomes similar to those of *M. musculus*; but the other had conspicuously large biarmed X chromosomes, a conspicuous Y, and a variable number of biarmed autosomes. Matthey sent all the skin and skull specimens to F. Petter, a mammalogist in Paris, to see if Petter could sort them out. Using minor morphological differences (which other taxonomists had considered mere individual variations within a species), Petter found two types of animals. His data matched the karyological observations of Matthey. Therefore, these two investigators (Matthey and Petter, 1968) concluded that these mice represented two sympatric species: (*M. booduga* with 40 telocentrics and *M. dunni,* the one with many biarmed autosomes and large sex chromosomes).

Nevertheless, cytogenetic data collected in the prebanding era were still highly empirical. No one was aware of the fact that the amount and the distribution of constitutive heterochromatin is so variable, even among individuals of the same population. No one even suspected that entire chromosome arms could be made of heterochromatin. Therefore, Matthey interpreted the *booduga–dunni* differences as the results of chromosome rearrangements. Frances Arrighi and I made the same kind of misinterpretations when we analyzed 19 species of *Peromyscus* representing all subgenera and found that all of them had a diploid number of 48 but the number of chromosome arms varied from 56 (8 biarmed chromosomes) to 96 (all biarmed chromosomes). Without the knowledge of heterochromatin distribution, we considered the phenomenon of variable number of chromosome arms (without the concomitant reduction of diploid number) as the result of reciprocal translocations and pericentric inversions (Hsu and Arrighi, 1968). This concept was corrected later when the C-band distribution was known (Pathak, *et al.*, 1973; Markvong *et al.*, 1975). In many cases, short arms may be made of heterochromatin and the euchromatic arms remain essentially unchanged.

The banding and other techniques took mammalian cytogenetics out of its slum period as much as they did human cytogenetics. A number of significant contributions were made within a short time span. Some of these are summarized in the following sections.

1. The Genomic Size The diploid number of mammals ranges from as low as 6 in the Indian muntjac (Wurster and Benirschke, 1970) to as high as 92 in *Anatomys leander* (Gardner, 1971). Superficially, this difference gives an impression that the genomic size of mammals is quite variable, but this impression may be deceptive. Species with higher diploid numbers always have short chromosomes and species with low diploid numbers always have long chromosomes. Ohno *et al.* (1964) measured the total length of the metaphase chromosomes of several mammalian species with different diploid numbers (from 12 to 60) and found that the values were very close. This means that the basic amount of genetic material is similar, but it is packaged differently. More packages (chromosomes) mean smaller packages and fewer packages mean larger packages. Measurements of DNA content per cell (both by the spectromicrophotometric method and by the newer flow microfluorometic method) also suggest that the amount of DNA is similar among various mammalian species.

Nevertheless, exceptional cases do occur. We have mentioned previously that in some species the amount of constitutive heterochromatin is considerably more than that of their relatives. For example, in *Peromyscus,* one species may possess 30% more DNA per cell than another, and all the excessive DNA is made of highly repetitive DNA or heterochromatin (Deaven *et al.*, 1977). In the ground squirrels (Figure 23.1), the DNA content per cell of *Ammospermophilus* may double that of its close relative *Spermophilus* (Mascarello and Mazrimas, 1977). Therefore, the conclusion that the mammalian genomic size is constant should apply to euchromatic material only.

Whether there is a optimal diploid chromosome number for a given amount of DNA is still a question. The diploid number of the majority of mammalian species is in the thirties, forties, and fifties, but within each taxon, one can always find exceptions. For example, most artiodactyls (deer, cattle, etc.) have high diploid numbers (sixties to seventies), but the lowest diploid number, 6, is also found in this order.

2. Conservatism in Arrangement of Genetic Material

Even though the banded metaphase chromosomes are not as refined as the banded polytene chromosomes, they are still useful for comparison between closely related species. It came as a surprise to us when Dean Stock, constructing C- and G-banded karyotypes of the African green monkey ($2n = 60$) and the rhesus monkey ($2n = 42$) as a routine exercise, discovered that many chromosomes of these two species have similar banding patterns.

When James Mascarello was working on the chromosomes of the wood rats (*Neotoma*), he one day had a number of photographic prints on his desk waiting to be sorted. My colleague Sen Pathak saw these prints of G-banded metaphases and thought that Jim had made a mistake and printed some *Peromyscus* chromosomes with which Sen had worked for some time. This again indicated that the banding patterns of some chromosomes in different genera of the same family, or even different families of the same order, can be used to trace homology and therefore phylogenetic relationships (Mascarello *et al.*, 1974). In *Neotoma* alone, Jim found that the degree of banding homology parallels the taxonomic status of the species. Even between Cricetidae (*Neotoma, Peromyscus*) and between Cricetidae and Muridae (*Rattus, Mus*) some chromosomes still appear unchanged.

**3. The
Robertsonian
Translocation**

The Robertsonian translocation is by far the most prevalent form of karyotypic evolution in mammals. One can find evidence for the Robertsonian process in practically every taxonomic denomination. One of the best examples is still Matthey's classic study of the African pygmy mouse, genus *Leggada* (Matthey, 1965). In this species, Matthey found that the basic diploid number of 36 acrocentric chromosomes could reduce, in different populations, with a concomitant increase of biarmed chromosomes. The lowest diploid karyotype is 18 biarmed chromosomes.

The classic concept of the Robertsonian (centric) fusion is that a break occurs at the long arm of an acrocentric immediately distal to the centromere and another break occurs at the minute short arm of the second chromosome. An asymmetrical translocation would result in a biarmed chromosome and a bare centromere which is subsequently lost. This scheme reduces the diploid number but not the fundamental number (NF). But if the break in the long arm is some distance away from the centromere, then the centromeric fragment may survive. In such cases, the diploid number remains the same after the translocation but NF is actually increased. A good example is found in the spotted skunk *Spilagale putorius* (Hsu and Mead, 1968). The typical karyotype of this species is 64 chromosomes with five pairs of biarmed elements. In a subspecies found in Oregon, the animal also had 64 chromosomes but the karyotype contained seven pairs of biarmed elements plus two pairs of minute metacentrics. Apparently there had been two sets of Robertsonian translocations but both centric fragments were preserved. Probably the supernumerary minutes found in the populations of the least harvest mouse *Reithrodontomys minimus* (Shellhammer, 1967) are also the relics of Robertsonian translocations.

There are other possibilities for centric fusions. White (1973) postulates that breaks may occur on the short arms of both acrocentrics. Then a restitution will result in a biarmed chromosome with two centromeres. The two centromeres are situated next to each other so that they may function as one. He also considers the possibility of breaks within the centromeres which will fuse into a compound centromere. Putative evidence for such mechanisms exists.

Some cytogeneticists consider that "fission," that is, splitting a biarmed chromosome into two acrocentrics, may be a force in karyological evolution. When the karyotypes of two taxa reveal a Robertsonian difference, they do not in any way indicate which direction, fusion or fission, has taken place. However, available evidence at least suggests that fusion is more common than fission.

In the mouse genus *Mus*, there is no question that a karyotype of 40 telocentrics is the most basic form. Numerous species in Southeastern Asia (including the laboratory mice) possess this karyotype. We are also reasonably certain that the European mouse populations were introduced by human agency, and therefore have a relatively short history. Yet in the Alpine valleys, Alfred Gropp, Ernesto Capanna, and their associates found varying numbers of biarmed chromosomes in feral mice to suggest Robertsonian translocation (Gropp *et al.*, 1970; Capanna *et al.*, 1973). Undoubtedly, these biarmed chromosomes were generated rather recently by fusions. The tobacco mouse *Mus poschiavinus* of Switzerland is recognized as a separate species. It has seven sets of centric fusions, thus reducing the diploid number to 26 but retaining the same FN. Capanna's group found a more extreme case: a diploid number of 22 (nine sets of centric fusions). However, all these mice can hybridize with regular *M. musculus* and produce offspring. The F_1 generation is partially sterile because of disturbances in meiosis.

One may argue that the mouse karyotype has no biarmed chromosomes originally. Without a biarmed chromosome, the only possible chromosome arm change would be fusion. Now let us examine the human karyotype, the mammalian species most extensively analyzed by cytogeneticists. The human karyotype has 10 acrocentrics and 36 biarmed chromosomes. If fusion and fission are of equal frequency, then more human beings should have a diploid number of 47 with two additional acrocentrics to replace one biarmed chromosome than a diploid number of 45 with a fusion between two acrocentrics. The actual situation is the reverse. Carriers with D/D, D/G, and G/G translocations are common. Yet in only one report (Hansen, 1975) was a plausible case of fission of chromosome 7 recorded. We can at least conclude that fusion, instead of fission, plays a major role in karyological evolution in mammals.

4. Tandem Translocations

A Robertsonian fusion combines two acrocentric chromosomes into one biarmed chromosome without changing the fundamental number. Therefore, from one form with all acrocentric chromosomes, the maximum reduction of the diploid number by the Robertsonian process will be 50%. However, in many instances the difference in diploid number within a taxon goes beyond such a limit, for example, the muntjacs. In the microtine rodents, the diploid number ranges from 56 in *clethrionomys* to 18 in *Microtus oregoni*. Unless some genetic material is discarded in the process of evolution (a most unlikely possibility), tandem translocations should be the prime candidate for producing the wide spectrum of diploid numbers in mammals.

In the human population, a growing list of cases has been accumulated to document that total chromosome translocations can be achieved, in addition to centromere–centromere fusions, by fusions between centromere and telomere and between telomeres. These are known as tandem translocations. For example, the two X chromosomes were found to fuse, end-to-end, into a very long submetacentric. In one case (Therman *et al.*, 1974) G-banding suggested no significant loss of chromatin in either X chromosome, and C-banding showed a piece of heterochromatin located at the centromere and an additional piece at the C-band location of the other X. This unique chromosome appeared monocentric in most metaphases; but in a small percentage of cells, it may appear dicentric or monocentric with the sister chromatids bending toward each other at the second centromere as if there was a precocious separation of the centromeres. These observations practically eliminated the possibility of the deletion of a centromere. The alternate hypothesis is that one of the two centromeres became inactivated to avoid chromatin bridge formation. Such incidences have become too numerous to be ignored.

A tandem translocation with a concomitant inactivation of one centromere allows for the reduction of the diploid number as well as the fundamental number. When two biarmed chromosomes fuse in this fashion, the diploid number will be reduced by two, whereas the NF will be reduced by four. Repeated tandem translocations may bring more than two chromosomes into a single element and reduce a very high diploid number to a very low diploid number. In primates, definite evidence of tandem translocations has been found among the karyotypes of Hominidae and Pongidae (DeGrouchy, Lejeune, and others), and among Cercopithecidae.

5. Inversions

In polytene chromosomes, even very short inversions can be detected and the break points identified in heterozygotes; but in ordinary metaphase chromosomes, paracentric inversions cannot be detected. Even in banded chromosomes, a paracentric inversion can be identified only when the inverted segment is long enough to reverse the banding pattern. This means that numerous genes must be included. In the human population, a few cases have been found, but undoubtedly most paracentric inversions escape detection.

Pericentric inversion is, of course, much easier to recognize. A good number of pericentric inversions have been found in the human population, many of which involve the pericentric C-bands. In other mammals, unequivocal pericentric inversions have also been found, both within a species and between species. In *Peromyscus,* for example, definite evidence for pericentric inversions has been obtained when G-banded karyotypes of *P. crinitus* and *P. leucopus* were compared (Arrighi *et al.,* 1976). There are not sufficient data to provide an estimate regarding the importance of this type of rearrangement in phylogeny, but it is not difficult to conceive that rearrangements such as inversions and translocations contribute significantly to reproductive isolation and hence speciation.

6. Heterochromatin

We have repeatedly mentioned heterochromatin (Chapters 8 and 18) as a special type of chromatin which apparently carries no informational genes. In the majority of animals, constitutive heterochromatin (C-band) is centromeric or paracentromeric. However, in a number of species, heterochromatin may have a variety of distributional patterns and variable amounts. It may be interstitial, terminal, or forming entire chromosome arms. We have mentioned that in *Peromyscus* a variable number of chromosome arms may be completely heterochromatic, and the number of heterochromatic arms is specific for each species. In many whales, squirrel monkeys, and the antelope squirrels (genus *Ammospermophilus*), each chromosome has its own unique pattern of heterochromatin content and distribution (Figure 23.1).

The discovery of many cases of totally heterochromatic chromosome arms makes it necessary to revise the concept of fundamental number. In karyotypic changes of strictly the Robertsonian type, NF is still a simple

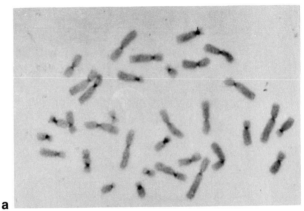

a

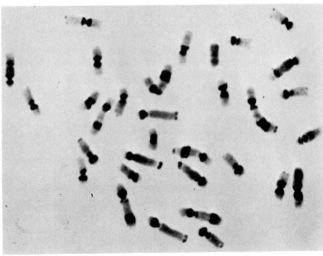

b

Figure 23.1. C-banded metaphases of two related genera of ground squirrels showing the dramatic differences in the amount and distribution of constitutive heterochromatin. **a,** *Spermophilus;* **b,** *Ammospermophilus* (Courtesy Dr. Sen Pathak.)

formula for crude comparison. But we have learned that species within a genus may differ in as many as 40 chromosome arms (*Peromyscus crinitus* and *P. eremicus*) without a major rearrangement. Thus unqualified application of NF for phylogenetic studies is to be avoided.

The exact role of heterochromatin in speciation is not yet apparent since the function of heterochromatin (or its basic component, highly repetitive DNA) is not yet understood. But the fact that some groups of animals possess a large amount of C-band material, whereas others do not certainly indicates that heterochromatin plays a role in karyotypic evolution. Identification of C-bands at least solves some perplexing phenomena, such as differences in genomic size, the change in fundamental number without pericentric inversions, etc. In comparing karyotypes of animal species, determination of the amount and the distribution of C-band is a must to avoid misinterpretations.

7. Supernumerary Chromosomes

Supernumerary chromosomes are not common in mammals, although such cases have been found. In the harvest mouse *Reithrodontomys minimus,* Howard Shellhammer (1967) found a variable number of supernumerary chromosomes, all being small, centric fragments (presumably heterochromatic). In the black rat *Rattus rattus* and in the Thai mouse *Mus shortridgei,* supernumerary chromosomes have also been found (Yosida and Sagai, 1975). The most dramatic example of supernumerary chromosomes in mammals was found by Peter Baverstock in the Australian rodent *Uromys* (Baverstock *et al.*, 1977). Up to 12 supernumerary chromosomes may be found in individuals, and there seems to be a relation between the number of the supernumeraries and the geographic distribution of the species. All the supernumeraries are heterochromatic as revealed by C-banding. Therefore, the situation is analogous to the B-chromosomes of maize. According to Peter's unpublished data, in collaboration with a molecular biologist, the species has a satellite DNA fraction which presumably is the principal material of these extra chromosomes. It appears that *Uromys* will be an excellent model system to study many aspects of repetitive DNA as well as the significance of heterochromatin in organic evolution.

The banding techniques have clarified a great deal of misconception based on information collected during the prebanding period and opened many avenues for more precise comparison of genetic makeup among related life forms. It is anticipated that more improvements will be made in the future, both in cytogenetics and in molecular biology, and students of evolutionary science will have many refined tools at their disposal to gain more knowledge of phylogeny and population structure.

24 The Future

We have witnessed repeatedly that new discoveries, by design or by accident, could trigger a flurry of research activities which continued until the usefulness of the discovery was exhausted. Such situations will undoubtedly recur, in cytogenetics as well as in other scientific disciplines. It is difficult, therefore, to predict what new directions we will be led in.

Nevertheless, one can at least speculate on the problems facing us in the field of cytogenetics. Probably the most pressing problem in chromosome research is the understanding of the molecular architecture of the chromosome. We know a great deal more about the molecular components of chromosomes than we did a decade or two ago, but we are still rather ignorant about their relationships and their interplays in various phases of cellular development. It is encouraging to hear from James Watson (1974), who organized the Cold Spring Harbor Symposium on Chromosome Structure and Function and wrote in the Foreword of the monograph:

The chromosomes of higher cells have long held a central role in the imagination of our better biologists. We know them, however, only at a semimolecular level that generates more frustration than satisfaction. . . . Now, however, there are very good reasons for believing that the essential structural features of chromosomes may be resolved over the next decade.

I sincerely hope that his forecast is right, because then I might still be around to see such a day. Previously, biologists studied molecules that make up chromosomes on the one side and the complex structure, the chromo-some, on the other. A glaring gap existed between the two ends. The gap has been narrowed recently by the active studies on the structure of chromatin, the nucleo-somes. The approach indeed may realize the dream of all cytologists from the last century to the present day: knowledge of the structure and function of chromo-somes.

Another urgent matter in cytogenetics lies in decipher-ing the eukaryote genome. Says Renato Dulbecco (1974) in the summary speech for still another Cold Spring Har-bor Symposium, this one on oncogenic viruses:

The [viral] genome will be sequenced, the structures of viral pro-teins will be determined, we will know what the transforming pro-tein is, and from that we may be able to infer how it works. It is difficult to predict, however, what will happen in the other side of transformation, the cellular side. Progress there will probably de-pend on breakthroughs in the methodology of studying the genome of eukaryote cells.

Renato's remarks pleased me because many vir-ologists use cultured cells only as a vehicle in which to grow viruses: They do not care to understand the cells because cell biology is not their business. A thorough understanding of the mammalian genome is imperative not only to oncogenic virus research but also to genetic research. It is sad that many cytogeneticists have not even made an attempt to learn a little about molecular biology, because if they had, they would have discovered the pleasure of deepening their field of inquiry. A few years ago, restriction endonucleases were known only to specialists; but now even laymen talk about recombinant DNA research, which utilizes these enzymes exten-sively. With a large number of competent investigators now engaged in genome analyses, achievements in this direction are expected.

Evolutionary studies using cytogenetic characteristics should gain more sophistication and momentum. Mor-phologic analyses of chromosomes, utilizing all the newly developed techniques, will continue, but gradually the emphasis should shift to the molecular aspects of cy-togenetics. Recently, DNA sequencing techniques have been developed after fractionation with gel elec-trophoresis. Perhaps one of these days biologists will be able to compare two species by reading their gene align-ments, base pair by base pair, along each chromosome.

Automated systems for cytogenetic work will probably be developed more vigorously than ever. In the 1960s, several teams started karyotype recording and analyses by machines and enjoyed a limited degree of success. The increasing demand for karyological makeup of patients in the clinics (especially amniocentesis) will require further improvements in the automation system. Mortimer Mendelsohn's CYDAC scanner connected with a computer, and other designs using similar or different principles, are expected to thrive for both routine and research work. The machines will probably never completely replace human eyes and hands, but they can take on the bulk of the workload, such as locating the mitotic figures, photography, measurements, DNA determinations, and many other chores.

The more recent invention of flow microfluorometry (FMF), which is capable of measuring DNA content per cell with an incredible speed, has myriad applications in cytogenetics and cell biology. Certainly, further improvements in resolution and applicability would make the instrument more and more useful.

Because of technical advances, cytogenetics and cell physiology of somatic cells have made far more advances during the last twenty-odd years than those of the germline cells. One of the difficulties in studying the biochemistry and molecular cytogenetics of mammalian gametogenesis is that there are many discrete stages of meiosis and spermiogenesis all of which are present in each seminiferas tubule. In somatic cells in culture, sophisticated methods for cell synchronization have been employed to analyze many events in each stage of the cell cycle, but synchronization techniques are not available for testicular elements. This is one area in which plant materials are superior (naturally synchronized meiosis) as Herbert Stern and Yasuo Hotta have ably exploited. The next best thing to cell synchronization is to fractionate and purify special categories of cells within a mixed cell population. Such techniques have slowly become established and predictably should accelerate our acquisition of knowledge of mammalian meiosis.

When asked by a journal to review the Cold Spring Harbor Symposium monograph on chromosome structure and function, I could not resist the temptation to plagiarize a one-liner from a cigarette advertisement as my opening remark: "Cytogenetics has come a long, long way, baby." From the way human and mammalian cytogenetics has developed, and from the trend that cytogenetics is intimately associated with other biological and medical disciplines, it is not difficult to predict that the growth rate of this field will remain exponential for a long time to come, but that it still has a long, long way to go.

Suggested Readings

Barker, B. E. (1969) Phytomitogens and lymphocyte blastogenesis. *In Vitro* **4**:64–79.

Benirschke, K. (Ed.) (1969) *Comparative Mammalian Cytogenetics*. Springer-Verlag, New York.

Chiarelli, A. B., and Capanna, E. (Eds.) (1973) *Cytotaxonomy and Vertebrate Evolution*. Academic Press, New York.

Ephrussi, B. (1972) *Hybridization of Somatic Cells*. Princeton Univ. Press, Princeton, New Jersey.

German, J. (Ed.) (1974) *Chromosomes and Cancer*. John Wiley & Sons, New York.

Harris, H. (1970) *Cell Fusion: The Dunham Lectures*. Oxford Univ. Press, London and New York.

Hsu, T. C. (1974) Longitudinal differentiation of chromosomes. *Ann. Rev. Genet.* **7**:153–176.

Hungerford, D. A. (1978) Some early studies of human chromosomes, 1879–1955. *Cytogenet. Cell Genet.* **20**:1–11.

Kampik, B. (1963) Darstellung der historischen Entwicklung unserer Kenntnisse uber die Gestalt und Zahl menschlicher Chromosomen, Unpublished doctoral dissertation, University of Münster.

Kato, H. (1977) Spontaneous and induced sister chromatid exchanges revealed by the BudR-labeling method. *Internat. Rev. Cytol.* **49**:55–97.

McKusick, V. A., and Ruddle, F. H. (1977) The status of the gene map of the human chromosomes. *Science* **196**:390–405.

Mitelman, F., and Levan, G. (1976) Clustering of aberrations to specific chromosomes in human neoplasms. II. A survey of 287 neoplasms. *Hereditas* **82**: 167–174.

Mittwoch, U. (1967) *Sex Chromosomes*. Academic Press, New York.

Ohno, S. (1967) Sex Chromosomes and Sex-Linked Genes. Springer-Verlag, Berlin.

Rao, P. N., and Johnson, R. T. (1974) Induction of chromosomes condensation in interphase cells. *Adv. Cell Molec. Biol.* **3:**135–189.

Riley, R., Bennett, M. D., and Flavell, R. B. (Eds.). (1977) A discussion of the mieotic process. *Phil. Trans. Roy. Soc. London,* B **277:**183–376.

Ringertz, N. R., and Sarage, R. E. (1976) *Cell Hybrids*. Academic Press, New York.

Ruddle, F. H., and Creagan, R. P. (1975) Parasexual approaches to the genetics of man. *Ann. Rev. Genet.* **9:**407–486.

Turpin, R., and Lejeune, J. (1965) *Les chromosomes humains*. Gauthier-Villars, Paris.

Wolff, S. (1977) Sister chromatid exchange. *Ann. Rev. Genet.* **11:**183–201.

References

Andres, A., and Navashin, M. (1936) Ein Beitrag zun morphologischen analyse der Chromosomen des Menschen. *Z. Zellforsch u mikroskop. Anat.* **24:** 411–426.

Arrighi, F. E., and Hsu, T. C. (1971) Localization of heterochromatin in human chromosomes. *Cytogenetics* **10:** 81–86.

Arrighi, F. E., Stock, A. D., and Pathak, S. (1976) Chromosomes of *Peromyscus* (Rodentia, Cricetidae). V. Evidence of pericentric inversions. In *Chromosomes Today,* Vol. 5 (P. L. Pearson and K. R. Lewis, Eds), John Wiley & Sons, New York, pp. 323–329.

Barr, M. L., and Bertram, E. G. (1949) A morphological distinction between neurons of the male and female, and the behavior or the nucleolar satellite during accelerated nucleoprotein synthesis. *Nature* **163:** 676–677.

Barski, G., Sorieul, S., and Cornefert, F. (1960) Production dans des cultures in vitro de deux souches cellulaires en association, de cellules de caractére "hybride". *C. R. Hebd. Seances Acad. Sci.* **251:** 1825–1827.

Barski, G., Sorieul, S., and Cornefert, F. (1961) "Hybrid" type cells in combined cultures of two different mammalian cell strains. *J. Nat. Cancer Inst.* **26:** 1269–1291.

Baverstock, P., Watt, C. H. S., and Hogarth, J. T. (1977) Chromosome evolution in Australian rodents. I. The Pseudomyinae, the Hydromyinae and the *Uromys/Melomys* group. *Chromosoma* **61:** 95–125.

Bayreuther, K. (1960) Chromosomes in primary neoplastic growth. *Nature* **186:** 6–9.

Belling, J. (1921) On counting chromosomes in pollen mother cells. *Am. Natural.* **55:** 573–574.

Benirschke, K., Malouf, N., Low, R. J., and Heck, H. (1965) Chromosome complement: Differences between *Equus Caballus* and *Equus przewalskii*, Poliakoff. *Science* **148**: 382–383.

Blakeslee, A. F., and Avery, A. G. (1937) Methods of inducing doubling of chromosomes in plants. *J. Hered.* **28**: 392–411.

Bleyer, A. (1934) Indications that mongoloid imbecility is a gametic mutation of degressive type. *Am. J. Dis. Child.* **47**: 342–348.

Böök, J., and Santesson, B. (1960) Malformation syndrome in man associated with triploidy (69 chromosomes). *Lancet* **1**: 858.

Boué, J. G., and Boué, A. (1966) Les aberrations chromosomiques dans les avortements spontantés humains. *Compt. Rend.* **263**: 2054–2058.

Boveri, T. (1914) *Zur Frage der Entwicklung maligner Tumoren*. Jena, Gustav Fischer-Verlag.

Capanna, E., Civitelli, M. V., and Cristaldi, M. (1973) Chromosomal polymorphism in an Alpine population of *Mus musculus* L. *Bull. Zool.* **40**: 379–383.

Carlson, P. S., Smith, H. H., and Dearing, R. D. (1972) Parasexual interspecific plant hybridization. *Proc. Nat. Acad. Sci. U.S.* **69**: 2292–2294.

Carr, D. H. (1967) Chromosome anomalies as a cause of spontaneous abortion. *Am. J. Obst. Gynecol.* **97**: 283–293.

Caspersson, T., Zech, L., Modest, E. J., Foley, G. E., Wagh, U. (1969) Chemical differentiation with fluorescent alkylating agents in *Vicia faba* metaphase chromosomes. *Exp. Cell Res.* **58**: 128–140.

Caspersson, T., Zech, L., and Johanson, C. (1970) Differential banding of alkylating fluorochromes in human chromosomes. *Exp. Cell Res.* **60**: 315–319.

Chen, T. R., and Ruddle, F. H. (1971) Karyotype analysis utilizing differentially stained constitutive heterochromatin of human and murine chromosomes. *Chromosoma* **34**: 51–72.

Chicago Conference. (1966) Standardization in human cytogenetics. Birth Defects Original Article Series 2. Pp. 1–21. National Foundation—March of Dimes, New York.

Christofinis, G. J. (1969) Chromosome and transplantation results of a human luekocyte cell line derived from a healthy individual. *Cancer* **24**: 649–651.

Chrustschoff, G. K., and Berlin, E. A. (1935) Cytological investigations on cultures of normal human blood. *J. Genet.* **31**: 243–261.

Chu, E. H. Y., and Giles, N. H. (1958) Comparative

chromosomal studies on mammalian cells in culture. I. The HeLa strain and its mutant clonal derivatives. *J. Nat. Cancer Inst.* **20**: 383–401.

Committee on Standardized genetic nomenclature for mice. (1972) Karyotype of the mouse, *Mus musculus. J. Hered.* **63**: 69–72.

Conger, A. D., and Fairchild, L. M. (1953) A quick freeze method for making smear slides. *Stain Technol.* **28**: 281–283.

Craig-Homes, A. P., and Shaw, M. W. (1971) Polymorphism of human constitutive heterochromatin. *Science* **174**: 702–704.

Cruciger, Q. V. J., Pathak, S., and Cailleau, R. (1976) Human breast carcinomas: Marker chromosomes involving lq in seven cases. *Cytogenet. and Cell Genet.* **17**: 231–235.

Davidson, R. G., Nitowsky, H. M., and Childs, B. (1963) Demonstration of two populations of cells in the human female heterozygous for glucose-6-phosphate dehydrogenase variants. *Proc. Nat. Acad. Sci. U.S.* **50**: 481–485.

Deaven, L. L., Vidal-Rioja, L, Jett, J. H., and Hsu, T. C. (1977) Chromosomes of *Peromyscus* (Rodentia, Cricetidae). VI. The genomic size. *Cytogenet. and Cell Genet.* **19**: 241–249.

Denver Report. (1960) A proposed standard system of nomenclature of human mitotic chromosomes. *J. Hered.* **51**: 221–241.

Dev. V. G., Tantravahi, R., Miller, D. A., and Miller, O. J. (1977) Nucleolus organizers in *Mus musculus* subspecies and in the Rag mouse cell line. *Genetics* **86**: 389–398.

DiPaolo, J. A., and Popescu, N. C. (1976) Relationships of chromosome changes to neoplastic cell transformation. *Am. J. Pathol.* **85**: 709–726.

DiPaolo, J. A., Popescu, N. C., and Nelson, R. L. (1973) Chromosomal banding patterns and *in vitro* transformation of Syrian hamster cells. *Cancer Res.* **33**: 3250–3258.

Dofuku, R., Biedler, J. L., Spengler, B. H., and Old, L. J. (1975) Trisomy of chromosome 15 in spontaneous leukemia of AKR mice. *Proc. Nat. Acad. Sci. U.S.* **72**: 1515–1517.

Drets, M. E., and Shaw, M. W. (1971) Specific banding patterns of human chromosomes. *Proc. Nat. Acad. Sci. U.S.* **68**: 2073–2077.

Dulbecco, R. (1974) Oncogenic viruses: The last twelve years. *Cold Spring Harbor Symposium on Quantitative Biology* **39**:

Dutrillaux, B., and Lejeune, J. (1971) Sur une nouvelle technique d'analyse du caryotype humain. *C. R. Acad. Sci. Paris* **272**: 2638–2640.

Edwards, J. H., Harnden, D. G., Cameron, A. H., Crosse, V. M., and Wolff, O. H. (1960) A new trisomic syndrome. *Lancet* **1**: 787–790.

Eiberg, H. (1974) New selective Giemsa technique for human chromosomes, cd staining. *Nature* **249:** 55.

Eigsti, O. H., and Dustin, P., Jr. (1955) *Colchicine: In Agriculture, Medicine, Biology, and Chemistry.* Iowa College Press, Ames.

Elsevier, S. M., and Ruddle, F. H. (1975) Location of gene coding for 18s and 28s ribosomal RNA within the genome of *Mus musculus. Chromosoma* **52:** 219–228.

Ephrussi, B., and Weiss, M. C. (1965) Interspecific hybridization of somatic cells. *Proc. Nat. Acad. Sci. U.S.* **53:** 1040–1042.

Farnes, P., Barker, B. E., Brownhill, L. E., and Fanger, H. (1964) Mitogenic activity of *phytolacca americana* (pokeweed). *Lancet* **2:** 1100–1101.

Flamm, W. G. (1972) Highly repetitive sequences of DNA in chromosomes. *Int. Rev. Cytol.* **32:** 1–51.

Ford, C. E., and Hamerton, J. L. (1956) The chromosomes of man. *Nature* **178:** 1020–1023.

Ford, C. E., Jones, K. W., Miller, O. J., Mittwoch, U., Penrose, L. S., Ridler, M., and Shapiro, A. (1959) The chromosomes in a patient showing both mongolism and the Klinefelter syndrome. *Lancet* **i:** 709–710.

Fredga, K. (1970) Unusual sex chromosome inheritance in mammals. *Phil. Trans. Royal Soc. London, B.* **259:** 15–36.

Funaki, K., Matsui, S., and Sasaki, M. (1975) Location of nucleolus organizers in animal and plant chromosomes by means of an improved N-banding technique. *Chromosoma* **49:** 357–370.

Gall, J. G., and Pardue, M. L. (1969) Formation and detection of RNA–DNA hybrid molecules in cytological preparations. *Proc. Nat. Acad. Sci. U.S.* **63:** 378–383.

Gall, J. G., Cohen, E. H., and Atherton, D. D. (1974) The satellite DNAs of *Drosophila virilis. Cold Spring Harbor Symposia on Quantitative Biology* **38:** 417–421.

Gardner, A. L. (1971) Karyotypes of two rodents from Peru, with a description of the highest diploid number recorded for a mammal. *Experientia* **27:** 1088–1089.

Gartler, S. M. (1967) Genetic markers as tracers in cell culture. *J. Nat. Cancer Inst. Monogr.* **26:** 167–195.

German, J. (1962) DNA synthesis in human chromosomes. *Trans. N.Y. Acad. Sci.* **24:** 395–407.

German, J. (1970) Studying human chromosomes today. *Am. Scientist* **58:** 182–201.

Gershon, D., and Sachs, L. (1963) Properties of a somatic hybrid between mouse cells with different genotypes. *Nature* **198:** 912–913.

Glade, P. R., Kasel, J. A., Moses, H. L., Wang-Peng, J., Hoffman, P. F., Kammermeyer, J. K., and Chessin, L. N. (1968) Infectious mononucleosis: Continuous suspension culture of peripheral blood lymphocytes. *Nature* **217:** 564–565.

Gropp, A., Tettenborn, U., and Lehmann, E. von (1970) Chromosomenvariation vom Robertsonischen Typusber der Tabakmaus *M. poschiavinus,* und ibren Hybriden mit der Laboratoriumsmaus. *Cytogenetics* **9:** 9–23.

Goodpasture, C., and Bloom, S. E. (1975) Visualization of nucleolar organizer regions in mammalian chromosomes using silver staining. *Chromosoma* **53:** 37–50.

Haldane, J. B. S. (1932) Genetic evidence for a cytological abnormality in man. *J. Genet.* **26:** 341–344.

Hamerton, J. L. (1971) *Human Cytogenetics,* Vol. 2. Academic Press, New York.

Hansen, S. (1975) A case of centric fission in man. *Humangenetik* **26:** 257–259.

Harnden, D. G., Miller, O. J., and Penrose, L. S. (1960) The Klinefelter-mongolism type of double aneuploidy. *Ann. Human Genet.* **24:** 165–169.

Harris, H., and Watkins, J. F. (1965) Hybrid cells derived from mouse and man: Artificial heterokaryons of mammalian cells from different species. *Nature* **205:** 640–646.

Hayflick, L., and Moorehead, P. S. (1961) The serial cultivation of human diploid cell stains. *Exp. Cell Res.* **25:** 585–621.

Henderson, A. S., Warburton, O., and Atwood, K. C. (1972) Location of ribosomal DNA in the human chromosome complement. *Proc. Nat. Acad. Sci. U.S.* **69:** 3394–3398.

Henderson, A. S., Eicher, E. M., Yu, M. T., and Atwood, K. C. (1974) The chromosomal location of ribosomal DNA in the mouse. *Chromosoma* **49:** 155–160.

Hilwig, I., and Gropp, A. (1972) Staining of constitutive heterochromatin in mammalian chromosomes with new fluorochrome. *Exp. Cell Res.* **75:** 122–126.

Hooper, E. T., and Musser, G. G. (1964) Notes on classification of the rodent genus *Peromyscus. Occas. Papers Mus. Zool. Univ. Michigan.* No. 635: 1–13.

Howard, A., and Pelc, S. R. (1953) Synthesis of deoxyribonucleic acid in normal and irradiated cells and its relation in chromosome breakage. *Heredity* (suppl.) **6:** 261–273.

Howell, W. M., Denton, T. E., and Diamond, J. R. (1975) Differential staining of the satellite regions of human acrocentric chromosomes. *Experientia* **31:** 260–262.

Hsu, T. C. (1952) Mammalian chromosomes *in vitro.* I. The Karyotype of Man. *J. Hered.* **43:** 172.

Hsu, T. C. (1961) Chromosomal evolution in cell populations. *Int. Rev. Cytol.* **12:** 1–68.

Hsu, T. C. (1965) Genetic cytology. In *Cells and Tissues in Culture* (E. N. Willmer, Ed.). Academic Press, New York. Pp. 297–361.

Hsu, T. C., and Arrighi, F. E. (1968) Chromosomes of *Peromyscus* (Rodentia, Cricetidae) I. Evolutionary trends in twenty species. *Genetics* **7:** 417–446.

Hsu, T. C. and Cooper, J. E. K. (1974) On diploid cell lines. *J. Nat. Cancer Inst.* **53:** 1431–1436.

Hsu, T. C., and Mead, R. A. (1968) Mechanisms of chromosomal changes in mammalian speciation. In *Comparative Mammalian Cytogenetics* (K. Benirschke, Ed.). Springer-Verlag, New York. Pp. 8–17.

Hsu, T. C., and Pomerat, C. M. (1953) Mammalian Chromosomes *in vitro*. II. A method for spreading the chromosomes of cells in tissue culture. *J. Hered.* **44:** 23–29.

Hsu, T. C., and Somers, C. E. (1961) Effect of 5-bromodeoxyuridine on mammalian chromosomes. *Proc. Nat. Acad. Sci. U.S.* **47:** 396–403.

Hsu, T. C., and Zenzes, M. T. (1964) Mammalian chromosomes *in vitro*. XVII. Idiogram of the Chinese hamster. *J. Nat. Cancer Inst.* **32:** 857–869.

Hsu, T. C., Arrighi, F. E., Klevecz, R. R., and Brinkley, B. R. (1965) The nucleoli in mitotic divisions of mammalian cells *in vitro*. *J. Cell Biol.* **26:** 539–553.

Hughes, A. (1952) Some effects of abnormal tonicity on dividing cells in chick tissue cultures. *Quart. J. Microscopical Sci.* **93:** 207–220.

Hungerford, D. A., and DiBeradino, M. (1958) Cytological effects of prefixation treatment. *J. Biophys. Biochem. Cytol.* **4:** 391–400.

Johnson, R. T., and Rao, P. N. (1970) Mammalian cell fusion: II. Induction of premature chromosome condensation in interphase nuclei. *Nature* **226:** 717–722.

Jones, K. W. (1970) Chromosomal and nuclear location of mouse satellite DNA of individual cells. *Nature* **225:** 912–915.

Kemp, T. (1929) Uber das Verhalten der chromosomen in den somatischen Zellen des Menschen. *Z. für mikroskop. anat. Forsch.* **16:** 1–20.

Kit, S. (1961) Equilibrium sedimentation in density gradients of DNA preparations from animal tissues. *J. Mol. Biol.* **3:** 711–716.

Kottler, M. J. (1974) From 48 to 46: Cytological technique,

preconception, and the counting of the human chromosomes. *Bull Hist. Med.* **48:** 465–502.

LaCour, L. F. (1944) Mitosis and cell differentiation in the blood. *Proc. Roy. Soc. Edinburgh, B.* **62:** 73–85.

Latt, S. A. (1973) Microfluorometric detection of deoxyribonucleic acid replication in human metaphase chromosomes. *Proc. Nat. Acad. Sci. U.S.* **70:** 3395–3399.

Latt, S. A. (1976) Optical studies of metaphase chromosome organization. *Ann. Rev. Biophys. Bioengin.* **5:** 1–37.

Lejeune, J., Gautier, M., and Turpin, R. (1959) Etude des chromosomes somatiques de neuf enfant mongoliens. *Compt. Rend.* **248:** 1721–1722.

Lejeune, J., Lafoureade, J., Berger, R., Vailatte, J., Boeswillwald, M., Seringe, P., and Turpin, R. (1963) Trois cas de deletion partielle du bras court d'un chromosome 5. *Compt. Rend.* **257:** 3098–3102.

Levan, A. (1938) The effect of colchicine on root mitosis in Allium. *Heredity* **24:** 471–486.

Levan, A., and Hauschka, T. S. (1952) Chromosome numbers of three mouse ascites tumors. *Hereditas* **38:** 251–255.

Levan, G., and Levan, A. (1975) Specific chromosome changes in malignancy: Studies in rat sarcomas induced by two polycyclic hydrocarbons. *Hereditas* **79:** 161–198.

Levan, G., and Mitelman, F. (1975) Clustering of aberrations to specific chromosomes in human neoplasms. *Hereditas* **79:** 156–160.

Lima-de-Faria, A. (1959) Differential uptake of tritiated thymidine into hetero- and euchromatin in *Melanoplus* and *Secale. J. Biophys. Biochem. Cytol.* **6:** 457–466.

Littlefield, J. W. (1964) Selection of hybrids from matings of fibroblasts *in vitro* and their presumed recombinants. *Science* **145:** 709–710.

Lyon, M. F. (1961) Gene action in the X-chromosome of the mouse. *Nature* **190:** 372–373.

McIntosh, A. J., and Sharman, G. B. (1953) The chromosomes of some species of marsupials. *J. Morph.* **93:** 509–532.

Makino, S. (1951) *An Atlas of the Chromosome Numbers in Animals.* Iowa State College Press, Ames, Iowa.

Makino, S. (1975) *Human Chromosomes.* Igaku Shoin, Tokyo.

Makino, S., and Hsu, T. C. (1954) Mammalian chromosomes *in vitro.* IV. The Somatic complement of the Norway rat, *Rattus norvegicus. Cytologia* **19:** 23–28.

Makino, S., and Nishimura, I. (1952) Water-pretreatment squash technic: A new and simple practical method for the chromosome study of animals. *Stain Tech.* **27:** 1–7.

Manolov, G., and Manolova, Y. (1972) Marker band in one chromosome 14 from Burkitt lymphomas. *Nature* **237:** 33–34.

Mark, J. (1977) Chromosomal abnormalities and their specificity in human neoplasms. An assessment of recent observations by banding techniques. *Adv. Cancer Res.* **24:** 165–222.

Markvong, A., Marshall, J. T., Jr., Pathak, S., and Hsu, T. C. (1975) The karyotypes of *M. fulvidiventris* and *M. dunni*. *Cytogenet. Cell Genet.* **14:** 116–125.

Mascarello, J. T., and Mazrimas, J. A. (1977) Chromosomes of antelope squirrels (Genus *Ammospermophilus*): A systematic banding analysis of four species with unusual constitutive heterochromatin. *Chromosoma* **64:** 207–217.

Mascarello, J. T., Stock, A. D., and Pathak, S. (1974) Conservatism in the arrangement of genetic material in rodents. *J. Mammal.* **55:** 695–704.

Matsui, S., and Sasaki, M. (1973) Differential staining of nucleolus organizers in mammalian chromosomes. *Nature* **246:** 148–150.

Matthey, R. (1945) L'evolution de la formule chromosomiate chez les vertebres. *Experientia* **1:** 78–86.

Matthey, R. (1949) Les chromosomes de vertebres. Lausanne, F. Rouge.

Matthey, R. (1951) Chromosomes de Muridae. *Experientia* **7:** 340–341.

Matthey, R. (1958) Un nouvelle type de determination chromosomique du sexe chez mammiferes *Ellobius lutescens* th. et *Microtus (Chilotus) oregoni* Bach. (Murides–Microtines). *Experientia* **14:** 7; 240.

Matthey, R. (1965) Cytogenetic mechanisms and speciations of mammals. *In Vitro* **1:** 1–11.

Matthey, R., and Petter, F. (1968) Existence de deux especes distincts, l'une chromosomiquement polymorphe, chez de *Mus* indiens du groupe *booduga*. *Rev. Suisse Zool.* **75:** 451–498.

Michel, W. (1937) Über die esperimentelle Fusion pflanzlicher Protoplasten. *Arch. Exp. Zellforsch. Besonders Gewebezuecht.* **20:** 230–252.

Miller, O. J., and Erlanger, B. F. (1975). Immunochemical probes of human chromosome organization. *Pathobiology Annual* (H. L. Iochim, Ed.) Appleton, New York. Pp. 71–103.

Miller, D. A., and Miller, O. J. (1972) Chromosome mapping in the mouse. *Science* **178:** 949–955.

Mittwoch, U. (1952) The chromosome complement in a Mongolian imbecile. *Ann. Engenics* **17**: 37.

Morishima, A., Grumbach, M. M., and Taylor, J. H. (1962) Asynchronous duplication of human chromosomes and the origin of sex chromatin. *Proc. Nat. Acad. Sci. U.S.* **48**: 756–763.

Moore, G. E., Gerner, R. E., and Franklin, H. A. (1967) Culture of normal human leukocytes. *J.A.M.A.* **199**: 519–524.

Moorhead, P. S., Nowell, P. C., Mellman, W. J., Battips, D. M., and Hungerford, D. A. (1960) Chromosome preparations of leucocytes cultured from human peripheral blood. *Exp. Cell Res.* **20**: 613–616.

Nichols, W. W., Levan, A., Aula, P., and Norrby, E. (1965) Chromosome damage associated with the measles virus *in vitro*. *Hereditas* **54**: 101–118.

Nowell, P. C. (1960) Phytohemagglutinin: An initiator of mitosis in cultures of normal human leukocytes. *Cancer Res.* **20**: 462–466.

Nowell, P. C. (1974) Diagnostic and prognostic value of chromosome studies in cancer. *Ann. Clin. Lab. Sci.* **4**: 234–240.

Nowell, P. C., and Hungerford, D. A. (1960) A minute chromosome in human chronic granulocytic leukemia. *Science* **132**: 1497.

Ohno, S., Becak, W., and Becak, M. L. (1964) X-autosome ratio and the behavior patterns of individual X-chromosomes in placental mammals. *Chromosoma* **15**: 14–30.

Ohno, S., Jainchill, J., and Stenius, C. (1963) The creeping vole (*Microtus oregoni*) as a gonosomic mosaic. I. The 0Y/XY constitution of the male. *Cytogenetics* **2**: 232–239.

Ohno, S., Stenius, C., and Christian, L. (1966) The X0 as the normal female of the creeping vole (*Microtus oregoni*). In *Chromosomes Today* (C. D. Darlington and K. R. Lewis, Eds.) Oliver & Boyd, Edinburgh and London. Pp. 182–187.

Okada, Y. (1958) The fusion of Ehrlich's tumor cells caused by HVJ virus *in vitro*. *Biken's J.* **1**: 103–110.

Okada, Y. (1962) Analysis of giant polynuclear cell formation caused by HVJ virus from Ehrlich's ascites tumor cells. I. Microscopic observation of giant polynuclear cell formation. *Exp. Cell Res.* **26**: 98–107.

Okada, Y., and Murayama, F. (1965) Multinucleated giant cell formation by fusion between cells of two different strains. *Exp. Cell Res.* **40**: 154–158.

Painter, T. S. (1922) The spermatogenesis of man. *Anat. Res.* **23**: 129.

Painter, T. S. (1923) Studies in mammalian spermatogenesis. II. The spermatogenesis of man. *J. Exp. Zool.* **37:** 291–336.

Pardue, M. L., and Gall, J. G. (1970) Chromosomal localization of mouse satellite DNA. *Science* **168:** 1356–1358.

Paris Conference. (1971) Standardization in human cytogenetics. Birth Defects Original Article Series 8: 1–46. The National Foundation—March of Dimes, New York.

Paris Conference Supplement. (1975) Birth Defects Original Article Series 11: 1–36. The National Foundation—March of Dimes, New York.

Pätau, K. (1961) Chromosome identification and the Denver report. *Lancet:* 933–934.

Pätau, K. A., Smith, D. W., Therman, E. M., Inhorn, S. L., and Wagner, H. P. (1960) Multiple congenital anomaly caused by an extra autosome. *Lancet* **1:** 790–793.

Pathak, S., Hsu, T. C., and Arrighi, F. E. (1973) Chromosomes of *Peromyscus* (Rodentia, Cricetidae) IV. The role of heterochromatin in karyotypic evolution. *Cytogenet. Cell Genet.* **12:** 315–326.

Patil, S. R., Merrick, S., and Lubs, H. A. (1971) Identification of each human chromosome with a modified Giemsa stain. *Science* **173:** 821–822.

Patton, J. L., and Gardner, A. L. (1971) Parallel evolution of multiple sex-chromosome system in the Phyllostomatid bats, *Carollia* and *Choeroniscus*. *Experientia* **27:** 105–106.

Patton, J. L. and Hsu, T. C. (1967) Chromosomes of the golden mouse, *Peromyscus (Ochrotomys) nuttalli* (Harlan). *J. Mammal.* **48:** 637–639.

Pearson, P. L., Bobrow, M., and Vosa, C. G. (1970) Technique for identifying Y chromosomes in human interphase nuclei. *Nature* **226:** 78–80.

Power, J. B., Cummins, S. E., and Cocking, E. C. (1970) Fusion of isolated plant protoplasts. *Z. Pflanzenphysiol.* **69:** 287–289.

Puck, T. T., Marcus, P. I., and Cieciura, S. J. (1956) Clonal growth of mammalian cells *in vitro*. Growth characteristics of colonies from single HeLa cells with and without a "feeder" layer. *J. Exp. Med.* **108:** 273–284.

Röhme, D. (1975) Evidence suggesting chromosome continuity during the S-phase of Indian muntjac cells. *Hereditas* **80:** 145–149.

Rothfels, K. H., and Siminovitch, L. (1958) An air-drying technique for flattening chromosomes in mammalian cells grown *in vitro*. *Stain Technol.* **38:** 73–77.

Rothfels, K. H., Axelrad, A. A., Siminovitch, L., McCulloch, E. A., and Parker, R. C. (1959) The origin of altered cell lines from mouse, monkey, and man, as indicated by chromosome and transplantation studies. *Can. Cancer Conf.* **3:** 189–213.

Rowley, J. D. (1973) A new consistent chromosomal abnormality in chronic myelogenous leukemia identified by quinacrine fluorescence and Giemsa staining. *Nature* **243:** 290–293.

Russell, L. B. (1961) Genetics of mammalian sex chromosomes. *Science* **133:** 1795–1803.

Sande, J. H. van de, Lin, C. C., and Jorgenson, K. F. (1977) Reverse banding on chromosomes produced by a guanine-cytosine specific DNA binding antibiotic: Olivomycin. *Science* **195:** 400–402.

Saunders, G. F. (1974) Human repetitions DNA. *Adv. Biol. Med. Phys.* **15:** 19–46.

Schweizer, D. (1977) R-banding produced by DNase I digestion of chromomycin-stained chromosomes. *Chromosoma* **64:** 117–124.

Seabright, M. (1971) A rapid banding technique for human chromosomes. *Lancet* **ii:** 971–972.

Sharman, G. B., McIntosh, A. H., and Barber, H. N. (1950) Multiple sex-chromosomes in the marsupials. *Nature* **166:** 996.

Shaw, M. W., and Krooth, R. S. (1964) The chromosome of the Tasmanian rat-kangaroo (*Potorous tridactylis apicalis*). *Cytogenetics* **3:** 19–33.

Shellhammer, H. S. (1967) Cytotaxonomic studies of the harvest mice of the San Francisco Bay Region. *J. Mammal.* **48:** 549–556.

Slifer, E. H. (1934) Insect development: VI. The behavior of grasshopper embryos in anisotonic, balanced salt solution. *J. Exp. Zool.* **67:** 137–157.

Smith, D. W., Pätau, K. A., Therman, E. M., and Inhorn, S. L. (1960) A new autosomal trisomy syndrome: Multiple congenital anomalies caused by an extra chromosome. *J. Pediat.* **57:** 338–345.

Sorieul, S., and Ephrussi, B. (1961) Karyological demonstration of hybridization of mammalian cells *in vitro*. *Nature* **190:** 653–654.

Sueoka, N. (1961) Variation and heterogenecity of base composition of deoxyribonucleic acids: A compilation of old and new data. *J. Mol. Biol.* **3:** 31–40.

Sumner, A. T., Evans, H. J., and Buckland, R. A. (1971) A new technique for distinguishing between human chromosomes. *Nature New Biol.* **232:** 31–32.

Sutton, W. D., and McCallum, M. (1972) Related satellite DNA's in the genus *Mus*. *J. Mol. Biol.* **71:** 633–656.

Szybalski, W. (1961) Properties and applications of halogenated deoxyribonucleic acid. In *The Molecular Basis of Neoplasia*. Univ. of Texas Press, Austin, Texas. Pp. 147–172.

Szybalski, W., Szybalska, E. H., and Ragni, G. (1962) Genetic studies with human cell lines. *Nat. Cancer Inst. Monogr.* **7**: 75–89.

Taylor, J. H. (1960) Asynchronous duplication of chromosomes in cultured cells of chinese hamster. *J. Biophys. Biochem. Cytol.* **7**: 455–464.

Taylor, J. H. (1958) Sister chromatid exchanges in tritium-labeled chromosomes. *Genetics* **43**: 515–529.

Taylor, J. H., Woods, P. S., and Hughes, W. L. (1957) The organization and duplication of chromosomes as revealed by autoradiographic studies, using tritium labeled thymidine. *Proc. Nat. Acad. Sci. U.S.* **43**: 122–128.

Therman, E., Sarto, G. E., and Pätau (1974) Apparently isocentric but functionally monocentrix X chromosome in man. *Am. J. Human Genet.* **26**: 83–92.

Therman, E., and Timonen, S. (1951) Inconsistency of the human somatic chromosome complement. *Hereditas* **37**: 266–279.

Timonen, S. (1950) Mitosis in normal endometrium and genital cancer. *Acta Obst. Gynecol. Scand.* **31** (Suppl. 2): 1–50.

Tjio, J. H., and Levan, A. (1956) The chromosome number of man. *Hereditas* **42**: 1–6.

Waardenburg, P. J. (1932) Das menschliche Auge und seine Erbanlangen. *Bibliog. Genet.* **7**.

Wang, H. C., and Fedoroff, S. (1972) Banding in human chromosomes treated with trypsin. *Nature New Biol.* **235**: 52–53.

Wang-Peng, J., Canellos, G. P., Carbone, P. P., and Tjio, J. H. (1968) Clinical implications of cytogenetic variants in chronic myelogenous leukemia (CML). *Blood* **32**: 755–766.

Watson, J. D. (1974) Foreword. *Cold Spring Harbor Symposium on Quantitative Biology* **38**: XV.

Weisblum, B., and DeHaseth, P. (1972) Quinacrine—a chromosome stain specific for deoxyadenylate-deoxythymidyte-rich regions in DNA. *Proc. Nat. Acad. Sci. U.S.* **69**: 629–632.

Weisblum, B., and Haenssler, E. (1974) Fluorometric properties of the bibenzimidazole derivative Hoechst 33258, a fluorescent probe specific for AT concentration in chromosomal DNA. *Chromosoma* **46**: 255–260.

Weiss, M. C., and Ephrussi, B. (1966) Studies of interspecific (rat × mouse) somatic hybrids. I. Isolation, growth and evolution of the karyotype. *Genetics* **54**: 1095–1109.

Weiss, M. C., and Green, H. (1967) Human–mouse hybrid cell lines containing partial complements of human chromosomes and functioning human genes. *Proc. Nat. Acad. Sci. U.S.* **58**: 1104–1111.

White, M. J. D. (1973) *Animal Cytology and Evolution.* Cambridge University Press, Cambridge, England.

Wilson, M. G., Towner, J. W., and Fujimoto, A. (1973) Retinoblastoma and D-chromosome deletions. *Am. J. Human Genet.* **25**: 57–61.

Winiwarter, H. von (1912) Etudes sur la spermatogenese humaine. *Arch. Biologie* **27**: 93; 147–149.

Wurster, D. H., and Benirschke, K. (1970) Indian Muntjac, *Muntiacus muntjac:* A deer with a low diploid number. *Science* **168**: 1364–1366.

Yerganian, G. (1957) The striped-back or Chinese hamster, *Cricetulus griseus. J. Nat. Cancer Inst.* **20**: 705–727.

Yerganian, G., and Nell, M. B. (1966) Hybridization of dwarf hamster cells by UV-inactivated Sendai virus. *Proc. Nat. Acad. Sci. U.S.* **55**: 1066–1073.

Yunis, G., and Yasmineh, W. G. (1971) Heterochromatin, satellite DNA and cell function. *Science* **174**: 1200–1209.

Yosida, T. H., and Sagai, T. (1975) Variation of C-bands in the chromosomes of several subspecies of *Rattus rattus. Chromosoma* **50**: 283–300.

Zakharov, A. F., and Egolina, N. A. (1972) Differential spiralization along mammalian mitotic chromosomes. I. BUdR-revealed differentiation in Chinese hamster chromosomes. *Chromosoma* **38**: 341–365.

Zech, L., Hagland, U., Nilsson, K., and Klein, G. (1976) Characteristic chromosome abnormalities in biopsies and lymphoid-cell lines from patients with Burkitt and non-Burkitt lymphomas. *Int. J. Cancer* **17**: 47–56.

Zimmerman, E. G., and Lee, M. R. (1968) Variation in chromosomes of the cotton rat, *Sigmoden hispidus. Chromosoma* **24**: 243–250.

Index